T0262756

European
Platforms
Initiative

ADVANCING
IoT PLATFORMS
INTEROPERABILITY

2018

River Publishers

Published, sold and distributed by:
River Publishers
Alsbjergvej 10
9260 Gistrup
Denmark

River Publishers
Lange Geer 44
2611 PW Delft
The Netherlands

Tel.: +45369953197
www.riverpublishers.com

ISBN: 978-87-7022-006-4 (Hard copy)
      978-87-7022-005-7 (ebook)

Disclaimer

The information in this document is provided as is and no guarantee or warranty is given that the information is fit for any particular purpose. The user thereof uses the information at its sole risk and liability.

The document reflects only the authors' views and the EC is not liable for any use that may be made of the information contained therein.

Table of contents

1

Executive summary

The IoT European Platforms Initiative (IoT-EPI) projects are address-ing the topic of Internet of Things and Platforms for Connected Smart Objects and aim to deliver an IoT extended into a web of platforms for connected devices and objects that supports smart environments, businesses, services and persons with dynamic and adaptive configu-ration capabilities. The specific areas of focus of the research activities are architectures and semantic interoperability, which reliably cover multiple use cases. The goal is to deliver dynamically-configured in-frastructure and integration platforms for connected smart objects covering multiple technologies and multiple intelligent artefacts. The IoT-EPI ecosystem has been created with the objective of increasing the impact of the IoT-related European research and innovation, in-cluding seven European promising projects on IoT platforms: AGILE, BIG IoT, INTER-IoT, VICINITY, SymbIoTe, bIoTope, and TagItSmart.

This white paper provides an insight regarding interoperability in the IoT platforms and ecosystems created and used by IoT-EPI. The scope of this document covers the interoperability aspects, challeng-es and approaches that cope with interoperability in the current ex-isting IoT platforms and presents some insights regarding the future of interoperability in this context. It presents possible solutions, and a possible IoT interoperability platform architecture.

Due to the critical and important role of interoperability in IoT systems, it is strongly related with many relevant topics and aspects in IoT: performance, compatibility, integration, ROI, market acceptance, development design, architecture.

The creation and development of this white paper has taken advantage from outcomes and available information from other activities in the framework of the IoT-EPI projects. In particular, important sources of information that complement the authors' research are the IoT-EPI platforms reports and the exchange of information among the projects during the work in task forces.

Regarding the output of this work, this white paper intends to be a useful source of information of interoperability in IoT platforms. This critical aspect in IoT systems has relevance and impact on the topics of each Task Force of the IoT-EPI (Innovation, Platform Interoperability, Community Building, Business Models, Educational Platforms, International Cooperation).

The research regarding interoperability architecture is useful for the current and future IoT platforms and IoT projects, as it provides deep awareness and valuable insights regarding the critical aspect of interoperability in them, as well as possible architecture solutions to the challenges that the achievement of platform interoperability involves. This information can be valuable in the development of new services, applications and businesses on top of IoT platforms.

Lack of platform interoperability causes major technologic and economic drawbacks such as impossibility to plug non-interoperable IoT devices into heterogeneous IoT platforms, impossibility to develop IoT applications exploiting multiple platforms, slowness of IoT technology introduction at a large-scale, discouragement in adopting IoT technology, vertical silos in IoT ecosystems and markets, increase of costs, scarce reusability of technical solutions, or user dissatisfaction.

In contrast, interoperability among platforms will provide numerous benefits such as new market opportunities, the disappearance of vertical silos, and vertically-oriented closed systems, architectures and application areas, to move towards open systems and platforms, and a major cooperation among platforms to offer better solutions to the consumer and the users. The cross-availability of services and data will allow current service providers to reach new markets with their services, but perhaps, more importantly, we expect new business opportunities to emerge from the ability to manage data from diverse sources to create innovative solutions.

2

IoT platforms landscape

2.1 Definitions

An IoT platform can be defined as an intelligent layer that connects the things to the network and that abstracts applications from the things with the goal to enable the development of services. IoT platforms achieve several main objectives such as flexibility (being able to deploy things in different contexts), usability (being able to make the user experience easy) and productivity (enabling service creation in order to improve efficiency, but also enabling new service development). An IoT platform facilitates communication, data flow, device management, and the functionality of applications. The goal is to build IoT applications within an IoT platform framework. An IoT platform allows applications to connect machines, devices, applications, and people to data and control centres [1].

IoT platforms' functionalities cover the digital value chain from sensors/actuators, hardware to connectivity, cloud and applications, as illustrated in Figure 2.1. Hardware connectivity platforms are used for connecting the edge devices and processing the data outside the datacentre (edge computing/fog computing), and program the devices to make decisions on the fly. The key benefits are security, interoperability, scalability and manageability by using advanced data management and analytics from sensor to datacentre. IoT software platforms include the integration of heterogeneous sensors/actuators; various

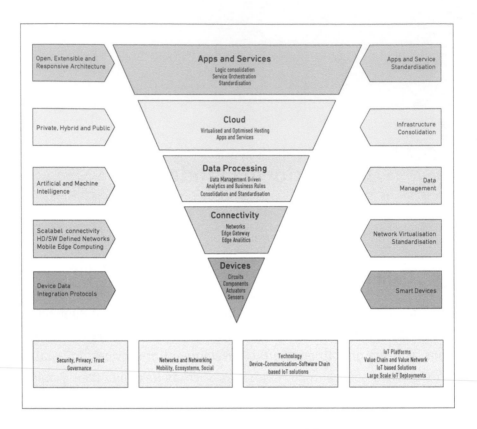

Figure 2.1. IoT Platforms covering the data value chain [1]

communication protocols abstract all those complexities and present developers with simple APIs to communicate with any sensor over any network.

IoT platforms also assist with data ingestion, storage, and analytics, so developers can focus on building applications and services, which is where the real value lies in IoT. Cloud-based IoT platforms are offered by cloud providers to support developers to build IoT solutions on their clouds. Infrastructure as a Service (IaaS) providers and Platform as a Service (PaaS) providers have solutions for IoT developers covering different application areas. PaaS solutions, abstract the underlying network, compute, and storage infrastructure, have focus on mobile and big data functionality, while moving to abstract edge devices (sensors/actuators) and adding features for data ingestion/processing and analytics services [1].

IoT platform definitions may differ in subtle and perhaps secondary details while overlapping on major features. One of the most succinct definitions of an IoT platform has been proposed by Gartner [2]. It defines the IoT platform as a software suite or a PaaS cloud offering that monitors, and may manage and control, various types of endpoints, often via applications end users build on the platform. It facilitates operations involving IoT endpoints and integration with enterprise resources. Platforms should be capable of:

- Provisioning and management of devices and their application software
- Data aggregation, integration, transformation, storage and management (often collectively referred to as "data digestion")
- Event processing (rule engine/orchestration/BPM)
- Customizing and building applications (SDK, app server, IDE and others)
- IoT data analysis and visualization
- Cybersecurity
- IoT device communications (network and/or Internet)
- Adapter (API hub, gateway software but also to the application on endpoint)
- User interfaces for both end users and developers

IoT platforms facilitate communication, data flow, device management, and the functionality of applications. A platform is not the application itself, although many applications can be built entirely within an IoT platform framework. It links machines, devices, applications, and people to data and control centres. It is not confined to brick-and-mortar central command; ideally, it can be accessed and managed from many different locus points. It employs better, quicker search engines and data storage systems with the capacity and sophistication to handle volume far beyond what has brought industry to the present moment [3]. Most of its elements are cloud-based and running on wireless connectivity, which may be established via third-party providers, application programming interfaces (APIs), cellular capabilities, or -most likely- a combination of these technologies.

Through dashboards, APIs, data engines, and algorithms, a platform enables elements and sectors of a business network to connect, monitor, and communicate with each other with far greater speed and flexibility than we have yet seen. Data from an ever-expanding ecosystem can be collected, sorted, and harnessed entirely online. The platform also can enable data prioritization, a feature of critical importance at a time when machines, sensors, and other objects are beginning to generate new floods of information.

IoT platforms provide security features, scalability, and capacity for pulling in, storing, and analysing data. It may connect machines, people, applications, or all three. Like any intelligent network, it provides innate predictive qualities that use data for the purposes of maintenance and troubleshooting. The user interfaces are intuitive and extensible, allowing for the future development of application extensions and the necessary scalability to track an increasing number of connected devices, people, and data sources.

Essentially, an IoT platform allows for greater concentration of resources in value-added applications. Instead of requiring companies to focus on the lower levels of the technology stack, which are essential but not value-positive, attention can be paid to application development; a smarter, more integrated IoT ecosystem; and intelligent data generation.

Using IoT platforms applications are sent to market faster and with better support. Connectivity and data management - which historically have required huge investments in time and development costs - are "givens" on the IoT platform, as reliable as electricity generation, and just as liberating to users.

The root of the IoT is connectivity: more things, more people, and the matrix of connections that springs up between them. Yet, in less than a decade, the technology has moved far beyond this fundamental consideration. Where many companies may have believed it was advantageous to build out a platform internally, it is becoming clear that much of the technology stack can now be implemented with out-of-the box tools and effective engagement with vendors.

ThingWorx [4] defines an IoT platform as a suite of components that enable:

- Deployment of applications that monitor, manage, and control connected devices.
- Remote data collection from connected devices.
- Independent and secure connectivity between devices.
- Device/sensor management.
- Integration with third party systems.

IoT platforms exist independently between the hardware and the application layers of the IoT technology stack. The ideal platform will integrate with any connected device, blend in with device applications, and enable implementation of IoT features and functions into any device in the same way.

Link Labs [5] defines an IoT platform at a high level as "the support software that connects edge hardware, access points, and data networks to other parts of the value chain (which are generally the end-user applications). IoT platforms typically handle ongoing management tasks and data visualization, which allow users to automate their environment. You can think of these platforms as the middleman between the data collected at the edge and the user-facing SaaS or mobile application".

IoT platforms are often referred to as middleware solutions, which can collectively be referred to as the **value chain of IoT,** that are the "plumbing" of the IoT. Generally, an IoT or M2M solution is a mash-up of functions from multiple vendors, which include:

- Sensors or controllers.
- A gateway device to aggregate and transmit data back and forth to the data network.

- A communications network to send data.
- Software for analysing and translating data.
- The end application service, which creates much of the value.

3

IoT platforms interoperability concepts, approaches and principles

Achieving interoperability is one of the main objectives of the IoT. As the name Internet of Things already states, it is all about connecting things and make them easily accessible just like the Internet today. "Broadly speaking, interoperability can be defined as a measure of the degree to which diverse systems, organizations, and/or individuals are able to work together to achieve a common goal" [6]. However, interoperability is a complex thing and there are many aspects to it. In literature, there exists quite a lot of different classifications of these aspects of interoperability, often also called levels of interoperability. One of the most important classification of levels of interoperability for technical systems is called *Levels of Conceptual Interoperability Model* (LCIM) and is depicted in Figure 3.1. Although it was created in the context simulation theory it has a much broader applicability. It defines six levels of interoperability: technical, syntactic, semantic, pragmatic, dynamic and conceptual interoperability.

The European Interoperability Framework designed "to support the delivery of pan-European eGovernment services to citizens and enterprises" [8] defines only three levels: technical, semantic and organisational interoperability.

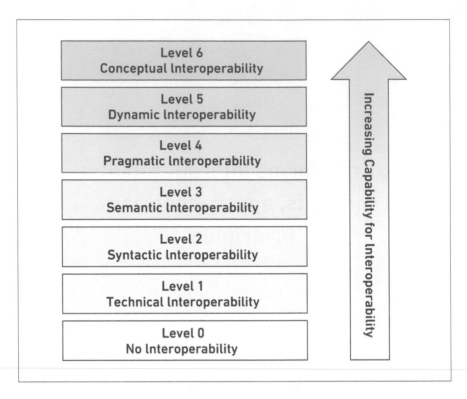

Figure 3.1. The Levels of Conceptual Interoperability Model [7].

A more IoT-specific classification is provided by ETSI and AIOTI and defines four levels [9]: technical, syntactic, semantic and organisational interoperability. In short, they are defined as follows.

- **Technical Interoperability:** usually associated with communication protocols and the infrastructure needed for those protocols to operate.
- **Syntactic Interoperability:** usually associated with data formats and encodings, e.g., XML, JSON and RDF.
- **Semantic Interoperability:** associated with the common understand of the meaning of the exchanged content (information).
- **Organisational Interoperability:** associated with the ability of organisations to effectively communicate and transfer infor-

mation even across different information systems, infrastructures or geographic regions and cultures.

As this is the most agreed-upon classification of interoperability levels within the IoT domain we follow it in this document.

3.1. IoT platforms interoperability challenges

3.1.1. Patterns of IoT interoperability

Achieving interoperability on the IoT, requires a closer look at interactions of the key components in IoT ecosystems. Before looking into the specifics of technical interoperability, syntactic interoperability, semantic interoperability, and organizational interoperability, we analyse here those interactions and we identify in Figure 3.2 six generic interoperability patterns for IoT ecosystems. This subsection is based on the material published in [10].

The "Cross Platform Access" pattern (Figure 3.2, I) is the basic characteristic of an interoperable IoT ecosystem. The goal of this pattern is to hide that an application or service accesses resources (information or functions) from different platforms through the same interface specification. The challenge of realizing this goal lays in allowing applications or services to discover platforms with relevant information, and enabling platforms that are potentially from different providers to have the same interface and use the same formats to communicate data.

The pattern "Cross Application Domain Access" (Figure 3.2, II) extends the "Cross Platform Access" pattern. The goal is that services/applications are able to access information and functions not only from different platforms, but also from platforms, which host information from multiple application domains. Thereby, it is crucial that semantic interoperability is given through well-defined and shared information models. A cross-domain application that accesses multiple IoT platforms, could e.g. air quality information and traffic data to provide green routing through a city.

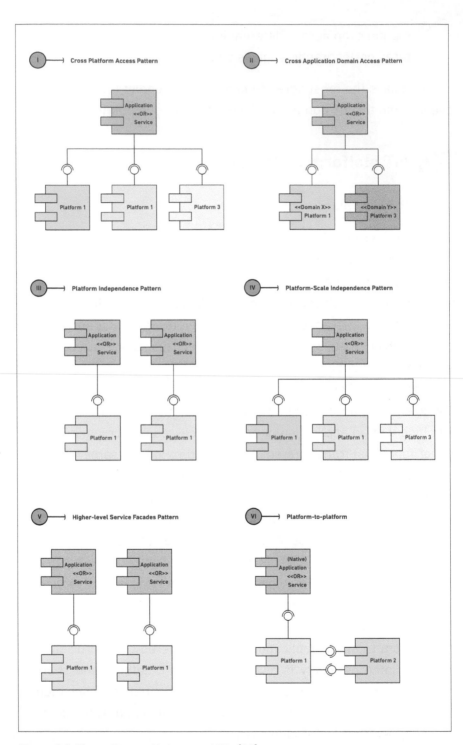

Figure 3.2. The patterns of interoperability [10].

The goal of the pattern "Platform Independence" (Figure 3.2, III) is to allow a single application or service to be used on top of different IoT platforms (e.g. in different regions). For example, these can be multiple deployments of a "smart parking" service used in two different geographic regions, which utilize different platforms with information about parking spots. This is especially challenging, when the sensors producing the IoT data are based on entirely different technology (e.g., radar-based parking spot observation, or counting in and outs of a parking lot).

The goal of the "Platform-Scale Independence" pattern (Figure 3.2, IV) is to hide different platform scales towards the connecting services and applications. The so called *server-level* platforms are platforms with many devices connected (e.g. a cloud platform), whereas *device-level* platforms grant direct access to attached sensors, and *fog-level* platforms are intermediaries such as edge gateways. A platform implementing this pattern has to hide its scale from applications and services accessing it.

Finally, the pattern "Higher-level Service Facades" (Figure 3.2, V) extends the interoperability requirements from platforms to higher-level services. Here, services are acting similar to platforms and also provide IoT offerings via a common interface. Such a service acts as a façade towards an IoT platform and use or process the IoT offerings of the platform to provide added value.

Once the above described patterns are implemented, they ensure ecosystem interoperability and allow an easy re-usage of IoT offerings from the various platforms of the ecosystem.

Organizational interoperability can be realized by *IoT platform federations* formed by multiple partnering institutions that collaborate by sharing IoT resources in locations originally out of their reach. This represents an additional horizontal integration that enables "Open networked" IoT business models according to the classification in [11].

We can define an IoT platform federation as an association of several platforms enabling their secure interoperation, collaboration and sharing of resources. Platforms can be enabled to perform collaborative sensing/actuation tasks and to interact directly so as to trade/share resources. A mechanism for defining and monitoring Service Level Agreements (SLAs) should be in place, while we can also envision the emergence of roaming *IoT devices*, where a device registered and managed by one platform is nomadic and can interact with resources in smart spaces managed by another federated platform (in a *visited smart space*). Federated platforms should of course control the terms under which a roaming IoT device is allowed to use resources in environments operated by visited platforms.

Platform-to-platform direct interactions enable existing (native) applications to use resources managed and operated by other federated platforms as if they were offered by a single platform, as shown in Figure 3.2.VI. This reduces the burden of interacting with multiple platforms from an application or a service, while platforms increase the portfolio of offered resources. For example, if Platform 2 offers to barter data produced by its static temperature sensors within a federation formed by platforms 1 and 2, this means that Platform 1 can use and offer temperature readings produced by those sensors as if Platform 1 was managing the devices.

Platform 1 offers in turn its temperature sensors located in another location to Platform 2. Therefore, an application or service can use the sensors from a single platform. Note that oneM2M has also identified such interaction between two platforms and tags the interface for platform interworking Mcc' (between two services providers).

3.1.2. Semantic Interoperability

Semantics, as seen in linguistics and philosophy, refers to the study of meaning, which means the relation of signifiers like words, symbols or signs and their denotation [12]. In computer science, the meaning of semantics is the same, but here the relations of signifiers and their denotation need to be understandable and process able by machines.

The most common way to achieve this is by using an ontology which is 'an explicit specification of a [shared] conceptualization' [6] and can be imaged like a formally defined information model. This is in line with the idea of the Semantic Web [13] introduced by Tim Berners-Lee in 2001, proposing the evolution of the Internet from a web of documents to a web of machine-readable and -understandable data. The corner stone technologies of the Semantic Web are the Resource Description Format[1] (RDF), a lightweight (meta data) data model for describing ontologies, and SPARQL Protocol and RDF Query Language[2] (SPARQL), which both are standardized by the World Wide Web Consortium (W3C).

To enable building new innovative applications which, make use of data from multiple IoT systems, spanning existing "IoT silos" these systems must not only be able to exchange raw data but also have a common understanding of its meaning. Unfortunately, even if today's IoT systems are willing to expose their data and resources to others, their semantically incompatible information models, offering different de-scriptions or even understandings of resources and operational proce-dures become an issue. Therefore, semantic interoperability is defined as "the ability of computer systems to exchange data with unambiguous, shared meaning" [14] is the key to "data exchange and service creation across large vertical applications", which is the next step of evolution of the IoT [15].

The challenge in achieving semantic interoperability is to find a way to provide this unambiguous, shared meaning of things, i.e. bridging the semantic gap between two (or more) platforms. Figure 3.4. shows possi-ble approaches to semantic interoperability and their classification into three types. The shown approaches are an extension to the ones present-ed in [16] and form a solution space where each approach to semantic interoperability can be located. The solution space in the original paper ranges from the *Core Information Model* approach where all communi-cating platforms agree on one shared model to the *Mapping between*

---

1.   https://www.w3.org/RDF/
2.   https://www.w3.org/TR/sparql11-overview/

Platform-Specific Information Models approach where each platform is free to use whatever information model they like and interoperability is only achieved through mapping between them. We define two more approaches extending this range called *Arbitrary Information Model (+ Domain-Specific Models)* which do not solve the problem of semantic interoperability in general but already provide the mechanisms needed to address semantic interoperability and therefore allow solving it on a different level.

We also provide a classification of these approaches according to three types of (semantic) interoperability: *by chance*, *by standardization* and *by mapping*. Semantic interoperability *by chance* means, that each platform is free to use whatever model they like but is only interoperable to any other platform that, by chance, uses the same model. Semantic Interoperability *by standardization* refers to the fact that there is some kind of agreement or standardization regarding at least parts of the used model and (semantic) interoperability *by mapping* refers to the fact that mapping logic is used to translate between different models.

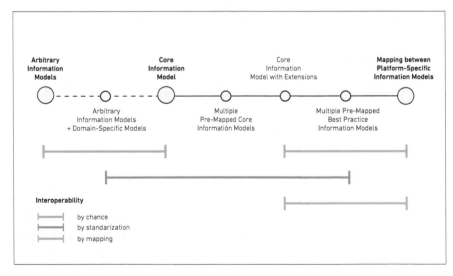

Figure 3.4. Possible approaches to semantic interoperability and their classification (based on [16]).

4

IoT-EPI Projects approaches addressing IoT platforms interoperability

---

4.1 IoT-EPI Platforms - Architectural mapping

The IoT platforms adoption is driven by factors such as economics that add cloud services and the development of partner ecosystems. In this context, device manufacturers provide built in solutions and models with the IoT SDKs to provide ease of use that allows the use of multiple portals and applications to get the IoT platforms and devices fully configured. The relationship with the service providers is increasingly important with the integration within the IoT suite and the various offerings from service providers.

The development of standardisation is accelerating in the area of device discovery to support ability for heterogeneous devices to communicate and interoperate. Standards are key to enabling interoperability, driving down costs and stimulating growth. However, standards processes are complex, take a long time to evolve and be adopted, and will still take some time to have mature, stable standards dominating, so suppliers and buyers are having to over-invest in multiple standards.

In this complex environment, the IoT-EPI projects are developing interoperability solutions that are addressing different layers in the IoT architecture and offer mechanisms for providing interoperability between different IoT platforms addressing various use cases and applications.

In the following paragraphs, we briefly discuss the mapping of the IoT architecture layers to the activities and solutions provided by the IoT-EPI projects.

AGILE builds a modular hardware and software gateway for the IoT focusing on the physical, network communication, processing, storage and application layers. The AGILE software modules are addressing functions such as device management, communication networks like area and sensor networks and solution for distributed storage. The project considers all the modules needed to provide a robust security management solution.

bIoTope provides an architecture and recommendations for the use of open standards and use case implementations that enable stakeholders to easily create new IoT systems and services and to rapidly harness available information using advanced Systems-of-Systems (SoS) capabilities for Connected Smart Objects. bIoTope also develops and provides standardised open APIs to enable interoperability. The project addresses all eight layers of the IoT architecture and validates the interoperability solutions in a cross-domain environment.

BIG IoT develops a generic, unified Web API for IoT platforms implemented. As part of the project, 8 partner IoT platforms are being integrated with the ecosystem plus several additional platforms are joining via the community building process. The project focuses on the upper layers of the IoT architecture by addressing the security management, APIs, service integration, external system services, applications, and the business enterprise.

INTER-IoT project addresses an open cross-layer framework, an associated methodology and tools to enable voluntary interoperability among heterogeneous IoT platforms by focusing on six layers of the IoT architecture with modules covering the QoS and device management, service integration, external system services, storage and virtualisation. The project addresses all network communication layer and the full security management suite.

symbloTe is providing an abstraction layer for a unified view on various IoT platforms and sensing/actuating resources. Applications can use symbloTe Core Services implementing a semantic IoT engine to find adequate resources offered by symbloTe-enabled platforms and subsequently access platform's virtual resources directly for data acquisition and actuation. The project focuses on seven layers of the IoT architecture from physical to application layer and proposes a full security management suite.

TagItSmart! offers a set of tools and enabling technologies that can be integrated into different IoT platforms using provided APIs to enable users across the value chain to fully exploit the power of condition-dependent functional codes to connect mass-market products with the digital world across multiple application sectors.

VICINITY focuses on a platform and ecosystem that provides "interoperability as a service" for infrastructures in the IoT and addresses the five-upper layer of the IoT architecture. The work considers the service integration, business logic, virtualisation, storage, APIs, tools, external system services, applications, data analytics and cloud services.

4.2 AGILE

The AGILE project aims to address technical and syntactic interoperability at hardware and software levels. On the hardware front the project designs hardware components extending the current state-of-the-art of available IoT gateway platforms with a twofold objective: to develop a so called "Maker's Gateway" by extending the capabilities of the most adopted and low-cost Raspberry Pi platform; and to develop a modular hardware gateway design for industrial purposes. For the Maker's Gateway, the project contributes a shield following the Raspberry HAT specification[3] to the Open Hardware community, extending the capabilities of the platform by two additional sockets for radio modules, with

---

3.   The Rapberry Pi B+ HAT (Hardware Attached in Top) specification is an extension hardware module specification for newer Raspberry Pi models, which has also been adapted by several other gateway class HW platform manufacturers.

several sensors including a GPS, and with further wired sensor connectivity options. The project also provides code that helps developers visualise and eventually use the features of these in a web-based visual application development environment called Node-Red. The ease of use for these tools enables people with no software / hardware competences to assemble their required solutions in a fast prototyping environment used also for testing purposes and before mass production.

On the software front the objectives of the AGILE project are to release open source code through the Eclipse Foundation to the community of IoT software developers / makers, helping them to easily configure their devices or gateways according to the platform environment these will be part of. The code is designed with gateway platform interoperability in mind, minimizing dependencies and thus supporting not just the two gateways hardware platform variants developed inside the project but also other platforms available on the market. To this end, Docker containerization (https://www.docker.com/) is used to separate software components from each other and from the underlying HW/OS. In fact, Docker is the leading containerisation technology for software containers, packages that contain software binary executables, runtimes and all related dependencies.

To further facilitate HW support, the project also contributes to the development of the Linux Yocto-based ResinOS operating system, specifically tuned for docker-based multi-container deployments adaptable to a large number of IoT gateway platforms.

Interoperability of platforms is therefore more easily fostered with the creation of an ecosystem of IoT applications that can be shared between users and developers leveraging existing initiatives like the Docker container ecosystem. Users are able to discover, install/manage and share components that have been developed for interoperability purposes in a secure way through the Docker app marketplace.

Figure 4.1 gives a more detailed view of what the AGILE Architecture looks like and which are the various modules that, run as ex-

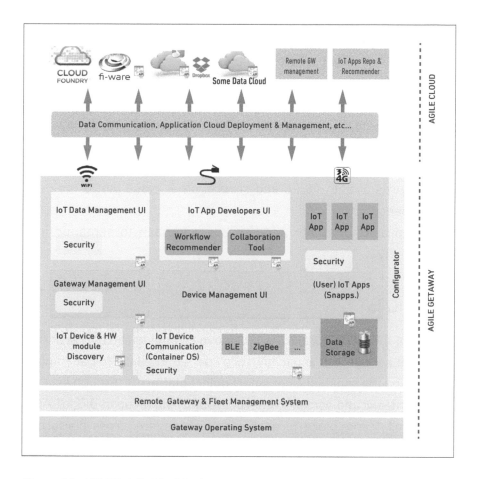

Figure 4.1. *AGILE Detailed Architecture*

ecutable images in a container platform, enable it to address interoperability. Worth highlighting for this purpose the IoT Device and HW module Discovery and IoT Device Communication (hardware interoperability) and the Device Management UI and IoT Data Management UI (software interoperability) that encompass all aspects of syntactic interoperability.

The top part of the picture shows how AGILE achieves <u>horizontal interoperability</u> of existing IoT platforms by allowing users to utilise external platforms for data management (e.g., Dropbox and Google-Drive), APIs for IoT devices (like wearable device APIs, home automation APIs, etc.) and application scalable deployment (e.g., support for CloudFoundry PaaS and integration of FIWARE enablers, Microsoft Azure, etc.). AGILE addresses these challenges at the software level by providing IoT platform specific modules for its Node-Red based visual application composition environment. With the categorization of Figure 3.2, this approach is similar to the "Cross Platform Access" pattern, with the notable difference that AGILE provides support in form of Node-Red modules, of which the application that uses features of both AGILE and several other cloud platforms can be composed.

The project does not address any approach to semantic interoperability.

4.3. BIG IoT

The goal of the **BIG IoT** project is to remove technological market entry barriers of service and application providers of the Internet of Things by exploiting the capabilities of **smart object platforms** through establishing *syntactic* and *semantic* **interoperability** via an open BIG IoT API and BIG IoT Marketplace to enable cross-standard, cross-platform, and cross-domain IoT services and applications. Thereby, the project is defining a comprehensive architecture [17] for IoT ecosystems including a solid security design [18] and service composition approach [19], while at the same time providing business models [20] that sustain the ecosystem.

4.3.1 The BIG IoT Architecture

This subsection is based on the material published in [10] and [17]. Below, the architectural approach and related interoperability aspects of the BIG IoT project are outlined. Figure 4.2 presents an overview of the BIG IoT architecture for IoT ecosystems. The architecture has been specifically designed to support all of the above-described patterns of interoperability (Section 4.1.1). The architecture is centred around a common set of interfaces, referred to as the BIG IoT API, that are

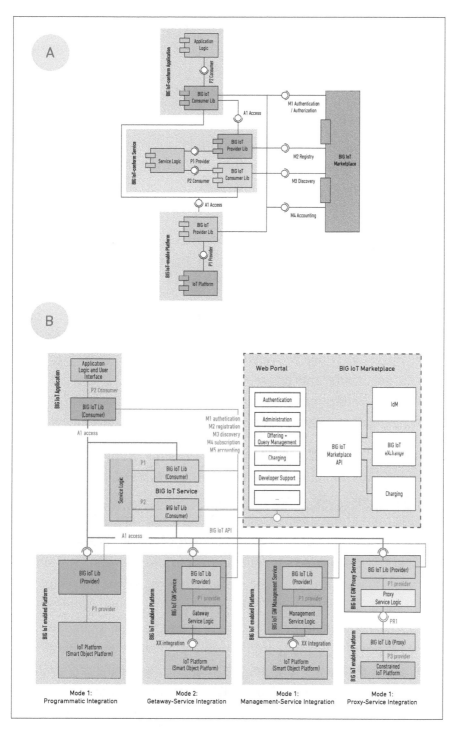

Figure 4.2. BIG IoT Architecture Overview A [10] and B [17] .

supported both by offering providers and consumers, as well as the marketplace, where resources are traded. These interfaces include the following basic interactions:

- M1: Authentication & authorization of offering providers and consumers
- M2: Registration of offerings (through offering providers)
- M3: Querying of offerings (through offering consumers)
- M4: Accounting of offering access
- A1: Access to offerings requested by offering consumers

These interfaces are the basis for enabling interoperability and realizing the patterns I – V (Section 4.1.1). Thereby, key challenges for realizing patterns II, III, and V are e.g.:

- The offering providers and consumers are from different application domains (II);
- The IoT offerings are hosted on different IoT platforms, e.g. located in different regions (III);
- The IoT offerings are on different provider systems, e.g. an IoT platform or a service (V).

Important in this figure is also the concepts of the BIG IoT Consumer and Provider Libs. For example, the Provider Lib implements the Register interface (M2) for uploading a description of an offering to the marketplace and offers the Access interface (A1) to provide the information to a consumer. The benefit of these libraries is that developers of platforms, services and applications are supported in advertising their offerings on the marketplace or use the marketplace to discover and access them. They only have to implement once the Provider (P1) or Consumer (P2) interface and can easily update the libraries in order to further comply in case of changes in the details of the underlying message formats and interactions. For the registration process via the Register interface (M2) a semantic description is used that is called an Offering Description and relies on the Resource Description Framework (RDF).

4.3.2. IoT platforms interoperability approach

A central goal of the BIG IoT architecture is to facilitate the integration of IoT platforms through the BIG IoT ecosystem. Both infrastructure as well as device-level platforms are targeted. From an architectural perspective, specifically considering the implementation of the BIG IoT API and integration with marketplaces, we have identified the following types of IoT platforms:

Type 1: Server Infrastructure or Cloud based IoT Platform assumed to be "always online" and anytime accessible by applications or services via the Internet.

Type 2: Device-level IoT Platform, hosted on devices that are _unconstrained_[4] with respect to communication, compute and memory resources assumed to be "always online" whereby connectivity and communication resources is assumed to be charged on a "flat-rate" plan.

Type 3: Device-level IoT Platform, hosted on devices that are _unconstrained_ with respect to communication, compute and memory resources, but are "not always online".

Type 4: Device-level IoT Platform, hosted on devices that are _unconstrained_-with respect to communication, compute and memory resources, but are connected to the Internet via a "pay-per-use" plan, Type 4 devices are often also of Type 3.

Type 5: Device-level IoT Platform, hosted on devices that are _constrained_[5] with respect to communication, compute and/or memory resources.

---

4. *Unconstrained* in this context means that the device will be able implement/use the BIG IoT API, which will be based on typical Web/Internet technologies (e.g. HTTP, WebSockets). An example of an unconstrained device is a Raspberry Pi. Also, other devices that are able to run Linux and support typical Web/Internet technologies are considered unconstrained in this context. A micro-controller based Sensor is not considered unconstrained.

5. *Constrained* in this context means with respect to the implementation of the BIG IoT API, which will be based on typical Web/Internet technologies (e.g. HTTP, WebSockets). An example of constrained devices are low-cost sensors, using a micro-controller. A Raspberry Pi is not considered a constrained device.

In the BIG IoT Project, overall 8 platforms are integrated by the partners. First, 6 cloud- or infrastructure-level platforms are part of the ecosystem: Bosch's Smart City platform, based on the Bosch IoT Suite[6] (**Type 1; Mode 1, Mode 2, Mode 4**), CSI's Smart Data[7] platform (**Type 1; Mode 3**), OpenIoT[8] (**Type 1; Mode 1, Mode 2**), VMZ's TIC[9] platform (**Type 1; Mode 1, Mode 2**), Siemens APM platform (**Type 1; Mode 2**), and WorldSensing[10] (**Type 1; Mode 1, Mode 2**). Further, there are 2 device-level platforms: Bosch's BEZIRK[11] platform (**Type 2, Type 3, Type 4; Mode 1, Mode 4**) and Econais' Wubby[12] platform (**Type 2, Type 3, Type 4, Type 5; Mode 4**).

Those now *BIG IoT-enabled platforms* are integrated via the BIG IoT API and BIG IoT Marketplace and currently being rolled out and tested in 3 European Pilot sites and applied in IoT scenarios for Smart Cities: Barcelona (BCN), Berlin/Wolfsburg (NG), and the region of Piedmont (Pied). Our Use Cases are divided in 9 clusters: Smart Parking (NG, BCN, Pied), Smart Traffic Management (BCN, Pied), Public Transport Optimization (NG, BCN), Healthy Bike Navigation (BCN, Pied), Smart Bike Sharing (NG, BCN, Pied), Incentive-based Green route Planning (BCN, Pied), Multi-Modal Route Optimizer (NG), Smart Charging (NG, BCN), Device-to-Device Communication (BCN), Smart Living (NG). In NG we use Bosch's Smart City platform, VMZ's TIC platform, Siemens APM Platform; in BCN are used OpenIoT, WorldSensing, Bosch's BEZIRK; Pied integrates CSI's Smart Data and Econais' Wubby platform.

4.3.3. Interoperability aspects

Solving interoperability issues related to these patterns requires the use of common information models, e.g., offered through Semantic

6.   https://www.bosch-si.com/products/bosch-iot-suite/platform-as-service/paas.html
7.   http://www.smartdatanet.it
8.   http://www.openiot.eu/
9.   https://viz.berlin.de/en/home
10.   http://www.worldsensing.com
11.   http://www.bezirk.com/platform.html
12.   http://www.wubby.io/

Web technologies. Those common information models need to allow the description of offerings, so that their consumers (e.g., services or applications) can work with them, even if they are from different domains or systems. The BIG IoT Core Model defines the core vocabulary required to create an Offering Description. The Offering Description relies on the work done within the W3C Web of Things (WoT) group, in particular, the Thing Description (TD) format [21] and going to be mapped also to W3C SSN/SOSA. The semantics are further enriched by domain independent and domain dependent models. A domain dependent model is used to semantically annotate the metadata, offering category and input/output data of an Offering Description. The BIG IoT semantic Application Domain Model is created using the BIG IoT semantic Core Model, domain independent and/ or domain models. This model establishes the relationship between the core model and domain model. Along with offering categorization and data modeling, the Domain Model also defines the vocabulary to semantically annotate the domain dependent features of an offering. For example, the class "mobility:ParkingSpace" can be used to annotate a parking space and its features such as parking space id, or parking space location.

For the semantic annotation of offering descriptions, e.g., defining the semantics of inputs, outputs and offering category, a well-defined vocabulary of domain terms is needed. This vocabulary should be widely shared and agreed upon so that all consumers and providers of IoT platforms can rely on it. Further, it should evolve in an open community process to allow active engagement by ecosystem stakeholders. We have selected schema.org as a basis for our domain model, as it provides a vendor-neutral, community-developed vocabulary for structured data.

The central pillar of the ecosystem is the Marketplace. Here, a Provider (e.g., an IoT platform) registers its resources by uploading an offering description (Section 3.2). To facilitate a provider in conforming with the BIG IoT API for offering its resources in the ecosystem, the Provider Library (Lib) can be utilized. It can be used to establish a gateway to the actual resources and implements the BIG IoT API and the various interactions and workflows (Section 3.4). The library authenticates the provider with the Marketplace and registers its offerings. It also offers a Web API to grant access to the resources.

Beyond these mechanisms for providing access to IoT platforms, the Marketplace enriches the ecosystem with the possibility of monetizing the consumption of resources. Therefore, both sides, the consumer and provider, report accounting data (e.g. number of resource records obtained/provided) back to the Marketplace. This is the basis for charging and billing, and the foundation for business opportunities around the ecosystem.

The above described IoT ecosystem is platform-scale independent. I.e., IoT platforms can operate either on cloud-level (e.g., server, data centre), on fog-level (e.g., gateway, cellular communication base station), or on device-level (e.g., Raspberry PI, smartphone). In the last case, the IoT platform can represent an IoT 'thing'. The BIG IoT API can be used independent of this scale of the platform.

4.3.4. Uniqueness and specific features of the approach

The BIG IoT approach comprises the following objectives: First, it openly defines the so-called BIG IoT API, a generic, Web-based application programming interface (API) for the adoption by smart object platforms.

Second, at the core of BIG IoT ecosystem stands an open marketplace. Once the BIG IoT API provides the building blocks for a syntactically and semantically interoperable IoT and inter-platform connectivity and exchange is enabled, the marketplace lays the foundation for an ecosystem of platforms and services offerings. The marketplace enables the advertisement, quality control, and monetization of applications and services developed on top of the BIG IoT API. A marketplace can be setup for specific domains, e.g., by stakeholders for the industry or energy domain, or by Smart Cities for the mobility or building domain.

Third, BIG IoT supports the development of applications and services by providing a dedicated software infrastructure. In addition, it provides functionality to discover and orchestrate services. This functionality is based on semantic service descriptions and an automated service chaining will be enabled through semantic reasoners. These functionalities facilitate the reusing of existing services.

Fourth, BIG IoT project engages with current standardization initiatives to receive input and further on, we will contribute to those standardization activities with the results elaborated in the project.

Finally, the core technologies related to BIG IoT API and Marketplace will be available as open source under Eclipse. The BIG IoT proposal has been approved and the Eclipse Bridge.IoT project has been created (https://projects.eclipse.org/proposals/eclipse-bridge.iot)

Used approach to semantic interoperability: Arbitrary Information Model and Domain-Specific Models

4.4. bloTope

A primary goal of bloTope is to enable companies to easily create new IoT systems and rapidly harness available information using advanced Systems-of-Systems (SoS) capabilities for Connected Smart Objects. This SoS approach signifies that all five patterns of interoperability in Figure 3.2 are implemented through open standards that can be combined and used together in different ways. The SoS approach taken by bloTope differs from most other proposed architectures for IoT systems in the sense that there is no layered architecture regarding the physical size or computational capabilities of the communicating systems. With this SoS approach, patterns I-III in Figure 3.2 are basic requirements that are implemented by default. Any system that implements the necessary IoT standards can communicate directly with any other system that implements and understands the same standards, as illustrated in Figure 4.3. This capability implements pattern *IV) "Platform-Scale Independence"* in Figure 3.2.

Systems that do not natively support the necessary IoT standards can join by what we call Wrappers in bloTope, i.e. software components that expose the services that should be published to the IoT
level using the appropriate standards (this corresponds to pattern *V)
"Higher-level Service Facades"* in Figure 3.2). This also signifies ensur-
ing that data and services that are not public remain non-accessible
to unauthorized parties.

In bloTope, "IoT standard" signifies a limited set of appropriate standards around the "waist" of the standards landscape as
illustrated in Figure 4.4. If the IoT is expected to become a similar success story as the World Wide Web (WWW), it will need a
similar foundation with a set of simple, well-defined, generic and

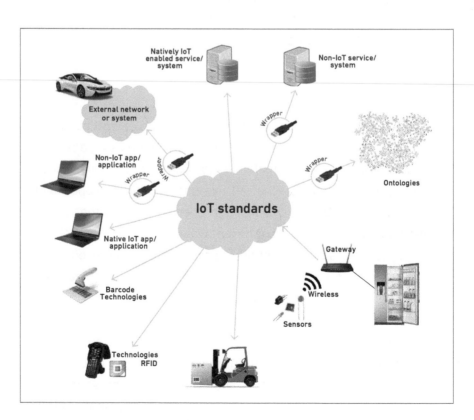

Figure 4.3. bloTope Systems of Systems type cross-connected (non-layered) connectivity.

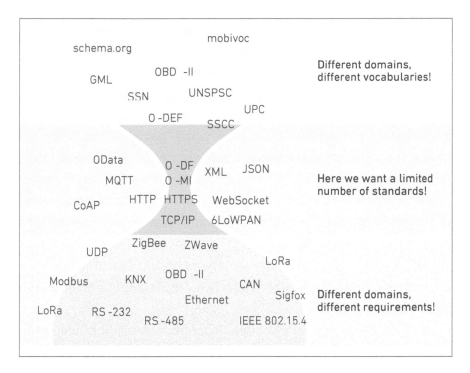

Figure 4.4. Illustration of bloTope standards landscape.

powerful standards. The WWW started with HTTP and HTML as the initial core standards. Over time supplementary functionality has been introduced by other standards that augment the capabilities of WWW applications but the fundamental building blocks and principles are still the same.

The core IoT standards used in bloTope are shown in bold. The other standards are used depending on the domain and requirements of different applications. When other standards (or proprietary protocols and formats) than the core IoT standards are used, bloTope uses a Wrapper (i.e., a *Service Facade*) for making them compliant with the core IoT standards.

bloTope takes full advantage of messaging standards developed and officially published by The Open Group, notably the Open Messaging Interface (O-MI) and Open Data Format (O-DF) standards[13]

---

13.   Formerly called QLM-MI and QLM-DF for historical reasons.

[22] [23]. Those standards emerged out of past EU FP6-FP7 projects, where real-life industrial applications required the collection and management of product instance-level information for many domains involving heavy and personal vehicles, household equipment, etc. Based on the needs of those real-life applications, and as no existing standards could be identified that would fulfil those requirements without extensive modification or extensions, the partner consortia started the specification of new IoT interoperability standards. O-MI mainly provides Technical Interoperability in the stack of Figure 2.1 with the necessary functionality needed for implementing generic IoT systems that is not provided by protocols such HTTP. O-MI can be used for implementing RESTful IoT information systems, also using other underlying protocols than HTTP as illustrated in Figure 4.5.

O-DF provides a generic content description model for Objects in the IoT that can be extended with more specific vocabularies (e.g., using domain-specific ontology vocabularies). O-DF currently uses XML as the underlying syntax but it provides a minimal, generic set of semantics for annotating IoT (and other) data. O-DF could be considered to bridge Level 2 (syntactic interoperability) and Level 3 (semantic interoperability) in Figure 4.5 because it provides a capability to reference external taxonomies, ontologies and vocabularies in a platform-, domain- and scale-agnostic way.

Figure 4.5 illustrates a bIoTope ecosystem, where the different IoT standard compliant systems (through Wrappers or not) are aware of each other's existence and the different data and services that they provide to each other[14]. When a new system needs to "join" a bIoTope compliant ecosystem, e.g. a car that arrives into a new city and needs to discover available services, bIoTope will provide discovery mechanisms for discovering relevant O-MI nodes. Certain nodes will imple-

---

14.   bIoTope assumes that different nodes can publish different data and services depending on the requesting node's identity, role, current context and other parameters. bIoTope also develops software components for the management of such access rights.

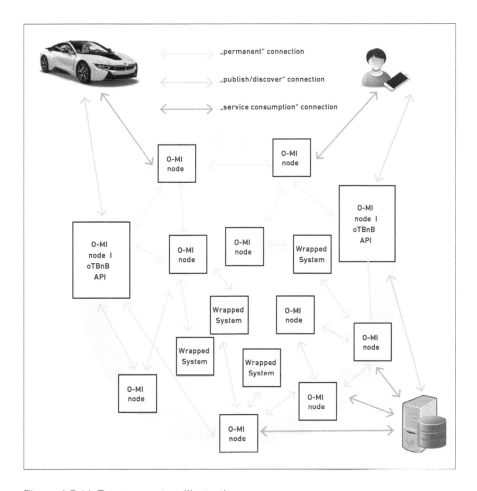

Figure 4.5. bioTope ecosystem illustration.

ment the bioTope IoTBnB API, which is a marketplace for IoT compliant services.

Figure 4.6 illustrates how different platforms, systems and services can publish the desired data and services using IoT standards. The Open Data Format (O-DF) provides means for annotating "any" IoT data or service using existing or new vocabularies, such as schema.org, SSN, GML, etc., as shown in Figure 4.4. It is even possible to use several different vocabularies within one O-DF structure (including proprietary ones) by linked data principles.

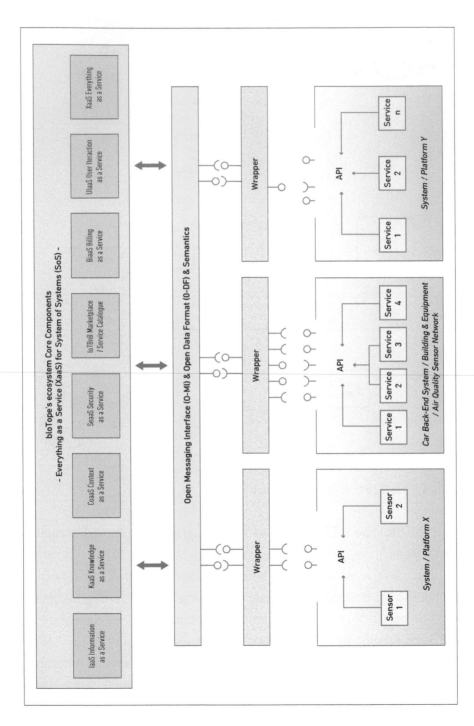

Figure 4.6. bIoTope architecture illustration.

The bIoTope architecture is heavily influenced by a Micro-Services Architecture (MSA) thinking, where different "small" standards can be used jointly in the best way depending on application requirements. MSA is a generic way of implementing pattern *V) "Higher-level Service Facades"* in Figure 3.2. Support for MSA is a fundamental design principle of O-MI and O-DF in the sense that services of "any" size can be published and consumed, no matter if it is a simple request for a current sensor value or a request for available parking places in proximity of a car's current location.

Such an MSA approach provides several advantages regarding system flexibility and lean software principles because new IoT standards-compliant services can be composed from smaller, existing services. It also provides a risk management advantages in the current competitive IoT standardization landscape because different standard-compliant software components and wrappers can be adapted to support new or "winning" standards without impacting the whole system architecture.

Whilst bIoTope should be understood as a highly flexible and dynamic ecosystem capable of seamlessly integrating arbitrary proprietary IoT platforms, it nevertheless builds upon a number of several core components presented in the top layer of Figure 4.6, which provide essential functionality. The basic viewpoint coordinates of from the architectural framework is presented in Figure 4.7. This figure adopts the conventions established, for example, by NIST[15] or IoT-A[16] with regards to the meanings of the cardinal directions North-South-East-West and set interaction patterns to human (north / west) and machines (south / east).

Nevertheless, the presented Services (XaaS) can be seen basically as chained micro services or functional blocks. Composed functional blocks result in a specific services, which as well can be part of an aggregation of several services. Figure 4.8 introduces a functional view onto those micro services, labelled as functional blocks. Basis for all

---

15.   https://www.nist.gov/sites/default/files/documents/itl/antd/Jeff_Voas.pdf

16.   http://www.meet-iot.eu/deliverables-IOTA/D1_3.pdf

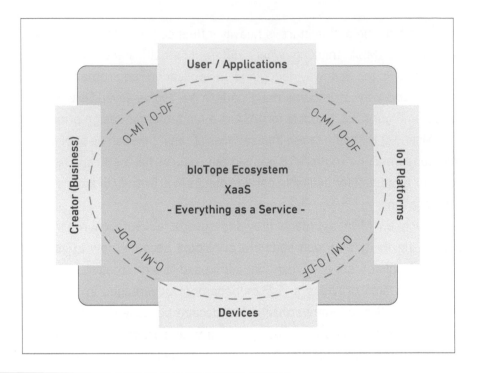

Figure 4.7. bloTope concept illustration.

functional blocks inside the ecosystem is the compliance to O-MI and O-DF. This compliance brings automatically the fundamental services for "publication" and "consumption" and their inherited sub-functions. For the interconnection of various services appearing in the ecosystem, the "marketplace / service catalogue" sets a cornerstone to interact as an intermediator between services. One possible instance of this services is represented by the IoTBnB[17] implementation. This specific implementation is as well supported by the Security & Privacy service that handles unauthorized actions. Inside of the ecosystem are other services (Visualization, Context Provisioning, Service Composition and RDF Integration & Semantics) that help to realize the stated core components / services on the top layer of Figure 4.6.

17.. http://iotbnb.jeremy-robert.fr/#/home

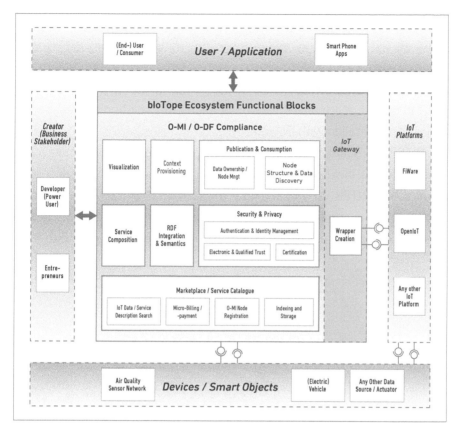

Figure 4.8. bloTope reference architecture.

Used approach to semantic interoperability: Arbitrary Information Model + Domain-Specific Models

4.5. INTER-IoT

4.5.1. Uniqueness and special features of the approach

INTER-IoT offers a layer-oriented solution for enabling seamless IoT platforms' interoperability. With this solution, different platforms can be interconnected and transparently interoperate among them at any specific layer or level (Device, Network, Middleware, Application and Data and Semantics). INTER-IoT is the first approach in providing universal semantic translation among platforms. It also provides a methodology (INTER-METH) for guiding and easing the implementation. INTER-IoT open framework (INTER-FW) offers a set of tools for interoperability at each specific layer which can be accessed through an API. Furthermore, INTER-IoT offers a virtualized version of each layer solution to facilitate a quick implementation with Docker.

INTER-IoT solution can be applied to any application domain and across domains in which there is a need for interconnection and/or interoperability. INTER-IoT will facilitate the formation of interoperable IoT ecosystems, make the design of IoT devices, smart objects or services easier to companies and developers, and support launching them to the

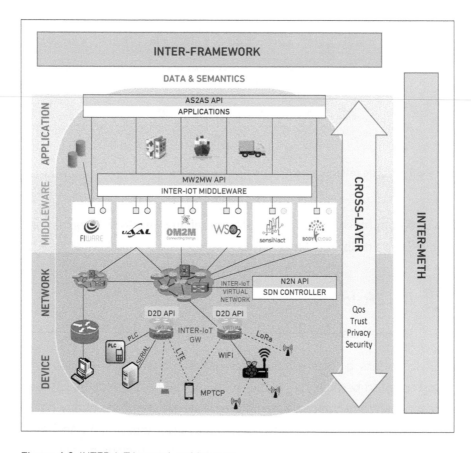

Figure 4.9. INTER-IoT layered architecture

market quickly to a broader client base. In the long term, the ability for applications to connect to and interact with heterogeneous smart objects will become a huge enabler for new products and services.

4.5.2.   INTER-IoT Architecture

INTER-IoT presents a novel layer-oriented solution for interoperability, to provide interoperability at any layer and across layers among different IoT systems and platforms. Contrary to a more general global approach, the INTER-IoT layered approach has a higher potential in order to provide interoperability. It facilitates a tight bidirectional integration, higher performance, complete modularity, high adaptability and flexibility, and presents increased reliability.

This layer-oriented solution is achieved through INTER-LAYER, several interoperability solutions dedicated to specific layers. Each interoperability infrastructure layer has a strong coupling with adjacent layers and provides an interface. Interfaces will be controlled by a meta-level framework to provide global interoperability. Every interoperability mechanism can be accessed through an API. The interoperability infrastructure layers can communicate and interoperate through the interfaces. This cross-layering allows to achieve a deeper and more complete integration.

4.5.3.   Layers and interoperability aspects

Device layer (D2D): At the device level, D2D solution will allow the seamless inclusion of novel IoT devices and their interoperation with already existing ones. D2D solution is a modular gateway that supports a vast range of protocols as well as raw forwarding. It is composed on a physical part that only handles network access and communication protocols, and a virtual part that handles all other gateway operations and services (**gw virtualization**). When connection is lost, the virtual part remains functional and is capable to answer the API and Middleware requests. The gateway follows a modular approach to allow the addition of optional service blocks to adapt to the specific case, allowing a fast growth of smart objects ecosystems.

Network layer (N2N): N2N solution enables seamless Network-to-Network interoperability, allowing transparent smart object mobility, and information routing support. It will also allow offloading and roaming, what implies the interconnection of gateways and platforms through the network. Interoperability is achieved through the creation of a virtual network, using SDN and NFV paradigms, with the support of the N2N API. The N2N solution will allow the design and implementation of fully interconnected ecosystems, and solve the smart object mobility problem.

Middleware layer (MW2MW): At the middleware level INTER-IoT solution will enable seamless resource discovery and management system for the IoT devices in heterogeneous IoT platforms. Interoperability at the middleware layer is achieved through the establishment of an abstraction layer and the attachment of IoT platforms to it. Different modules included at this level provide services to manage the virtual representation of the objects, creating the abstraction layer to access all their features and information. Those services are accessible through a general API. Interoperability at this layer will allow a global exploitation of smart objects in large scale multi-platform IoT systems.

Application and Services layer (AS2AS): INTER-IoT will enable the use of heterogeneous services among different IoT platforms. Our approach will allow the discovery, catalogue, use and even composition of services from different platforms. AS2AS will also provide an API as an integration toolbox to facilitate the development of new applications that integrate existing heterogeneous IoT services.

Semantics and Data layer (DS2DS): INTER-IoT guarantees a common interpretation of data and information among different IoT platforms and heterogeneous data sources that typically employ different data formats and ontologies, and are unable to directly share information among them. INTER-IoT DS2DS approach is the first solution that provides **universal semantic and syntactic interoperability** among heterogeneous IoT platforms. It is based on a novel approach, a semantic translation of IoT platforms' ontologies to/from a common Central Ontology that INTER-IoT employs, instead of direct platform-to-platform translations. This technique reduces dramatically the number of potential combinations of semantic translations required for universal semantic interoperability. INTER-IoT semantic interoperability tools work with any vocabulary, or ontology. INTER-IoT own modular Central Ontology, called GOIoTP, for all IoT platforms, devices and services, is available at http://docs.inter-iot. eu/ontology. Also, syntactic translators allow interoperability between different data formats, such as JSON, XML, and others. Although the pilot deployments of INTER-IoT realize the Core Information Model with Extensions approach to semantic interoperability, INTER-IoT supports any solutions between its pilot approach and Core Information Model.

Cross-Layer: Inter-IoT also guarantees non-functional aspects that must be present across all layers: trust, security, privacy, and quality of service (QoS). As well, INTER-IoT provides a virtualized version of the solution for each layer, to offer the possibility of a quick and easy deployment. Security is guaranteed inside each individual layer, and the external API access is securitized through encrypted communication, authentication and security tokens. INTER-IoT accomplishes the new European Data Privacy Law, and in the specific case of e-Health, in which information is highly sensitive, the Medical Device Regulation law.

Regarding the architectural interoperability patterns described in Section 3.1.1, INTER-IoT supports all six patterns of interoperability:

- Cross Platform Access, which is accomplished through AS2AS services or through MW2MW.
- Cross Application Domain Access, as far as INTER-IoT is domain-agnostic and has universal semantic interoperability by means of the DS2DS solution.
- Platform Independence, through AS2AS service composition and DS2DS semantic and syntactic translation.
- Platform-Scale Independence, by means of INTER-IoT AS2AS.
- Higher-level Service Facades, through INTER-IoT AS2AS services.
- Platform-to-Platform interaction, INTER-IoT main goal by design. It is achieved through D2D and/or MW2MW solutions.

All the aforementioned patterns are architectural. INTER-IoT has identified main patterns of interoperability from a different point of view or analysis: from the semantic point of view, regarding semantic interoperability and from the middleware interoperability point of view (related with syntactic and functional interoperability). INTER-IoT has its own macro and micropatterns that match with this approach.

Finally, regarding the three main types of interoperability (functional, syntactic, semantic), INTER-IoT enables all of them. Universal syntactic and semantic interoperability among any platforms with different data formats and ontologies is possible through the INTER-IoT DS2DS solution. Moreover, other INTER-IoT layers (D2D & N2N) can provide functional interoperability among smart elements, enabling connectivity to the network.

4.6. symbIoTe

The main goal of **symbIoTe (symbiosis of smart objects across IoT environments)** is to devise a **flexible and secure interoperability middleware** across IoT platforms facilitating rapid development of IoT applications across platforms, platform collaborations as well as dynamic and adaptive smart spaces. This is accomplished by 1) a **semantic IoT search engine** for connected (virtualized) smart objects (i.e., IoT resources) registered by platform providers, 2) **abstraction layer** for unified and secure usage of those resources across platforms, 3) **high-level, domain-specific APIs** ("Enablers") for rapid cross-platform application development, 4) **IoT platform federations**, i.e., associations between two platforms facilitating their secure interaction, collaboration and bartering of resources, 5) **dynamic and self-configurable** smart spaces offering interoperability for collocated devices and gateways, and 6) **secure interworking protocol** between the IoT platforms, gateways and smart devices. This supports SMEs and new entrants in the IoT market to build innovative IoT services within short development life cycles.

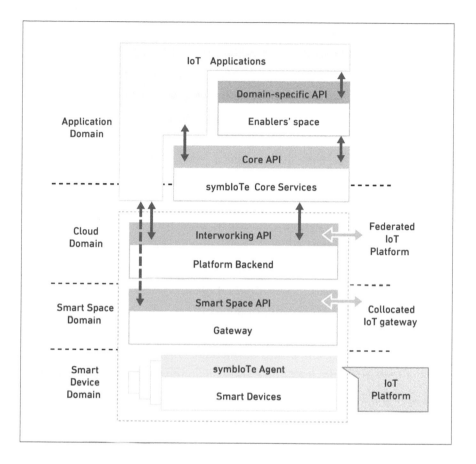

Figure 4.10. The symbloTe high-level architecture

4.6.1.   The symbloTe Architecture

The symbloTe architecture [24] is built around a layered IoT stack connecting various devices (sensors, actuators and IoT gateways) within
Smart Spaces with the Cloud. Smart Spaces share available local resources (connectivity, computing and storage), while platform services
running in the Cloud enable IoT Platform Federations and open the Interworking API[18] to third parties with flexible access control. The architecture comprises four layered domains, as depicted in Figure 4.10.

---

18.   Interworking Interface is a symbloTe defined interface which opens up platform resources
    as RESTful IoT Services in the Cloud Domain.

1) Application Domain enables platforms to register IoT devices which they want to advertise and make accessible via the Interworking API, while symbIoTe provides the means to search for IoT devices across platforms by means of its Core Services. We also build domain-specific back-end services (Enablers) which utilize the infrastructure provided by the underlying platforms to offer value-added services, e.g., data analytics on top of sensor data acquired from different platforms. Enablers ease the process of cross-platform and domain-specific application development (specifically for mobile and web IoT applications).

Cloud Domain provides a uniform and attribute-controlled access [25] to virtualized IoT devices exposed by platforms to third parties through an open API (Interworking API). In addition, it builds services for IoT Platform Federations enabling close platform collaboration and resource bartering, in accordance with platform-specific business goals and defined Service Level Agreements (SLAs).

Smart Space Domain offers services for discovery and interworking of collocated IoT devices and gateways in local spaces, while Smart Device Domain relates to the roaming capabilities of smart devices that maintain their identity while moving through different spaces.

4.6.2. IoT platforms interoperability approach

symbIoTe allows for flexible IoT platforms interoperability mechanisms (direct platform-to-platform interactions within **platform federations**, platform-to-platform interactions **within Smart Spaces**) which is achieved by an incremental deployment of symbIoTe functionality across the introduced architectural domains so that platform providers can choose an appropriate interoperability solution and integrate only selected features with their platforms. This in effect influences the level of platform interoperability and collaboration with other platforms within a symbIoTe-enabled ecosystem. The currently conceived solution has a minimal impact on existing IoT platforms as it requires mostly the development of a small interworking module. In addition symbIoTe is adding security layer offers a distributed, and effective mechanism for controlled access to resources in a federation of heterogeneous IoT platforms.

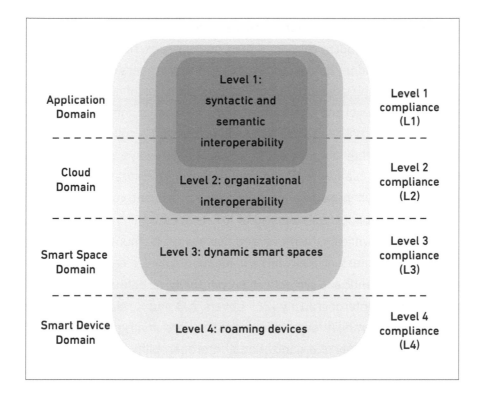

Figure 4.11. symbloTe Compliance Levels

4.6.3. Interoperability Aspects

The symbloTe approach defines **four interoperability levels,** as depicted in Figure 4.11. We also refer to them as **compliance levels,** when
considering them from the perspective of an IoT platform wanting to
become interoperable. In all four levels, interoperability is achieved by
offering a unified and secure way to advertise, find and consume IoT
resources, but in each level, a different interoperability scenario is enabled offering various degrees of details about the involved resources.
    Level 1 enables interactions between IoT applications and virtual-
ized IoT resources (**both sensors and actuators**), i.e., applications can
find and use resources across platforms through uniform interfaces.
In addition, platforms can integrate symbloTe components for flexible
attribute-based access control. **Level 2** allows IoT platforms to collab-

orate closely by forming **federations**. Federations can be considered as a closed and distributed version of the Core Services, i.e., the platforms can advertise and barter resources only to the members of the federation. **Level 3** enables **dynamic smart spaces**: adoption of new resources at the gateway level and direct interactions between symbloTe-enabled IoT devices (e.g. mobile devices and Arduino boards) in collocated smart spaces, even if they are connected to various gateways and managed by different IoT platforms. This enables resource migration between collocated and in proximity IoT gateways to prevent vendor lock-in. **Level 4** offers support for **roaming of IoT devices** which maintain their unique identity while on the move, and can always be found via the Core Services, regardless of their current location. Level 1 can be directly mapped to *semantic and syntactic interoperability*, as defined in the IERC Whitepaper on Interoperability [26]. Levels 2, 3 and 4 relate to *organizational interoperability*. symbloTe proposes here an original approach with finer granularity of organizational interoperability by placing specific interoperability concepts in either the Cloud or Smart Space.

4.6.4. Uniqueness and specific features of the approach

The first unique feature of symbloTe is its solution for semantic interoperability that follows the *Core Information Model with Extensions* approach presented in Section 3.1.2. The *Core Information Model* (CIM) is the central information model used to describe resources registered to symbloTe and is shared between all platforms. It is designed to be as abstract and minimalistic as possible and at the same time as specific as needed to create a minimal mutual understanding. This way, the CIM enables limited out-of-the-box interoperability between all platforms. To actually register resources with symbloTe, platforms use *Platform-Specific Information Models* (PIMs) that are extensions of the CIM including platform-specific terms. As each PIM is an extension of the CIM, basic interoperability is ensured, even between platforms using different PIMs. Such flexibility goes beyond the state of the art and is expected to lead to a high acceptance by platform owners, especially SMEs, as the process of adapting their platform's information model to symbloTe is significantly simplified while it offers the means to use their existing information models.

Furthermore, symbloTe develops a strategy to enable a higher-level of interoperability between platforms using different PIMs. It is based on a translation between different PIMs based on declarative mappings using SPARQL query re-writing and data translation. This concept has the potential to create high impact on semantic interoperability for IoT solutions since the process of information model standardization is frequently lengthy and cumbersome, while it also limits platform's flexibility.

The second unique feature relates to the implemented *semantic search engine for IoT resources* that is provided by the Core Services that let IoT application developers find adequate IoT resources for various cross-platform and cross-domain applications. The search engine uses and supports the listed symbloTe information models.

The third uniqueness relates to direct platform-to-platform interactions within **platform federations** for closer collaboration between platforms. Here symbloTe goes a step further to provide novel functionalities for enriching platform interactions: resource bartering, trust and SLA management, as well as support for smart device roaming across federated platforms.

The fourth uniqueness relates to the **security perspective.** symbloTe implements a flexible, distributed, and effective mechanism for controlled access to resources, both sensors and actuators, in a federation of heterogeneous IoT platforms. The conceived methodology grounds its roots in the powerful Attribute-Based Access Control mechanism. A specific access policy, defined as a combination of attributes (i.e., user's properties, roles, details) is assigned to each resource, while the access to that resource can be granted only to users in possession of a set of attributes matching the aforementioned policy.

The above listed features allow symbloTe to support all six patterns of interoperability presented in Section 3.1.1.

All six patterns of interoperability presented in Section 3.1.1 are supported by symbloTe and can be mapped to the above listed features. Pattern I, II and III are supported through the provided Core services (e.g.

search for resources), the Interworking API, support for PIMs and semantic mapping. Pattern IV (*Platform-Scale Independence*) fits to the four interoperability/compliance levels of symbIoTe; pattern V (*Higher-Level Service Facade*) matches the concept of Enablers and pattern VI (*Platform-to-Platform Direct Interactions*) the concept of federations as defined in symbIoTe.

symbIoTe is implementing an **Open Source** middleware prototype available at https://github.com/symbiote-h2020, following an **agile-like** approach. Developers from all consortium partners join forces in the implementation of the software components using the microservices architecture to fulfil the requirements of incremental feature deployment and scalability of the highly-distributed IoT ecosystem. Regarding **licensing**, the consortium has selected a "copyleft" license for the Core Services (LGPL-3.0), so that updates, bug fixes and new features are always given back to the Open Source Community. For the middleware components residing at the platforms' side the licensing is following the "non-copyleft" approach (the BSD 3-Clause is selected).

4.7. TagItSmart

A comprehensive view of TagItSmart architecture components is presented in Figure 4.12. TagItSmart is designed as a set loosely coupled components which can be integrated in and across different environments (IoT platforms). To this end, components providing specific TagItSmart functionality like smart tag encoding, decoding etc. come with their own APIs, while more generic IoT functions are reused from the underlying IoT platform used. This approach makes use of TagItSmart functionality rather straightforward in different settings and reduces the learning curve for developers.

From the functional point of view, the *User/Developer level* provides the front end functionality enabling access to different TagItSmart components. Additionally, at this level, the TagItSmart SDK enables integration of the smart tag decoding functionality into third party (smartphone) applications.

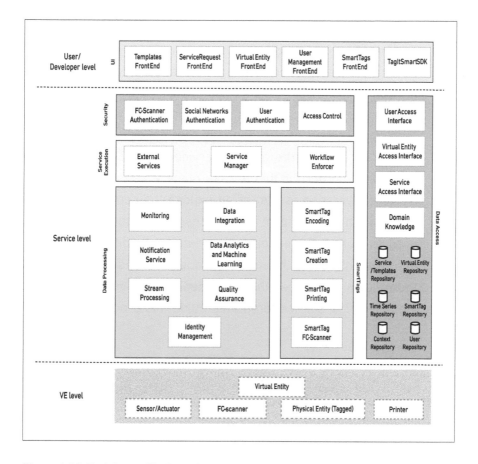

Figure 4.12. TagItSmart Platform Functional Architecture

At the **Service level**, we can find the following functional blocks:

- *Security* components deal with aspects such as authentication, authorisation and access control to the rest of the components. The security is addressed on several levels: the security of the smart tags is ensured with Ciphertext Policy Attribute Based Encryption (CP-ABE) by embedding policy contained in the ciphered tag in order to enforce access control policies associated with the different users; the security of the web platform and the APIs are ensured with the identity management based on credentials and authentication mechanism; and trustworthiness and data integrity hinge on the distributed ledger technology.

- *Service Execution* components include those that enable execution of services registered in the platform, as well as the
  service templates used to trigger dynamic creation of workflows. The overall process allows users of the platform to issue service requests that are then processed by the platform,
  resolved in terms of sub-services that need to be "glued together" to support the overall service request and are then
  executed. In order for this to take place, a series of repositories are leveraged to first of all expose the services that are
  available in the platform, so they can be discovered, configured, triggered and reused appropriately. In addition, a component that intercepts the service requests and is able to break
  them down to services that are needed are involved (Service
  Manager). The service templates guide this process of identi-

Table 4.1 Integration Strategies

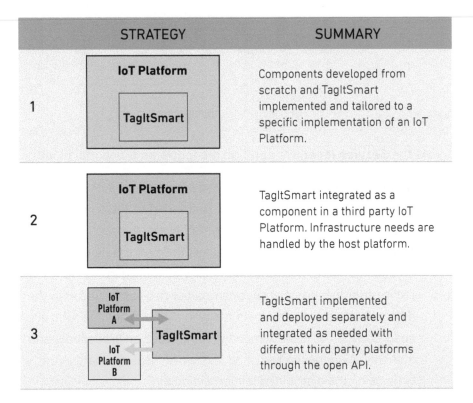

	STRATEGY	SUMMARY
1	**IoT Platform** / **TagItSmart**	Components developed from scratch and TagItSmart implemented and tailored to a specific implementation of an IoT Platform.
2	**IoT Platform** / **TagItSmart**	TagItSmart integrated as a component in a third party IoT Platform. Infrastructure needs are handled by the host platform.
3	**IoT Platform A** / **IoT Platform B** / **TagItSmart**	TagItSmart implemented and deployed separately and integrated as needed with different third party platforms through the open API.

fying what a service request needs in terms of sub-services as well user information. The latter both for making sure that only authorised users are allowed to access the platform and issue service requests as well as be served through the platform (e.g. an FC-scanner user to be able to use a decoding or stream processing component that is hosted in the TagItSmart platform) but also for identifying appropriate resources based on ownership, location and other availability criteria that need to be involved during a service request execution. Eventually the runtime execution of a service is undertaken by the Workflow Enforcer, which acts as a a "middle man" between components, passing information between components appropriately and handling errors so that other components can focus solely on their functional operation.

- *Data Processing* components provide additional functionality to handle and work with the data generated in the platform.
- *SmartTags* components facilitate integration, creation and scanning of the SmartTags.
- *Data Access* components provide the corresponding registries, semantic models and repositories on which TagItSmart operates.

At the **Virtual Entity** level, the actual representations of the objects that are part of the platform provide access to their data and defined actions based on the semantic models.

Taking the architecture described above as the starting point, here we describe how TagItSmart functionality can be implemented in practice:

Full Implementation of an IoT Platform with TagItSmart: this scenario builds up from scratch all the components defined in the architecture of TagItSmart the functionality is integrated in the same codebase. The infrastructure needed to support the platform is provisioned by the implementation provider.

Hosted in a third party IoT platform: this scenario integrates TagItSmart functionality in an already existing IoT platform. Components implementation might need some adaptation to comply with the existing platform, yet they should be interoperable by following the correspondent API. Infrastructure needs are set up by the third party IoT platform provider.

Shared and hosted externally: this scenario considers TagItSmart as a standalone set of services that expose an open API and that can be accessed from different platforms simultaneously. Deployment and infrastructure needs are handled independently from the IoT Platform.

The components defined in Figure 4.12. are the set of components needed to fulfil the requirements extracted from the different use cases, while enabling creation and lifecycle management of the SmartTags. Based on the chosen integration strategy, mapping of some of the components to real implementations and deployments in a specific IoT platform will be different.

Table 4.2. TagItSmart APIs example

OPERATION	API	DESCRIPTION
SmartTag activation	POST /product/set-smart-tag-date	This method is used to activate the SmartTag to be enable it use in the system.
SmartTag scan	POST /product/product-context	This method is used to POST the location where the product was scanned and returns all related context for the product including reviews and sensor states.
SmartTag recycle	POST /recycl3r/api/v1/ recycling_info	This method is used to get information about sorting and location for product recycling.

As stated above, TagItSmart components define a common set of API methods to guarantee interoperability and seamless integration, as well as enabling creation of the third-party applications on top of its components (Table 4.1).

4.8. VICINITY

The VICINITY project is built around the concept of connecting different IoT ecosystems through the VICINITY platform (interoperability as a service), which enables to interact with IoT objects from other different ecosystems as if they were their own. The interoperability services create an environment where value-added services can be deployed and processes make available cross-domain information.

In the presented figure, two separate ecosystems are presented: intelligent building and energy ecosystem. Each of these eco-

Figure 4.13. VICINITY Concept

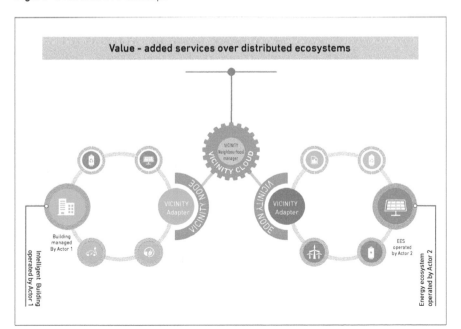

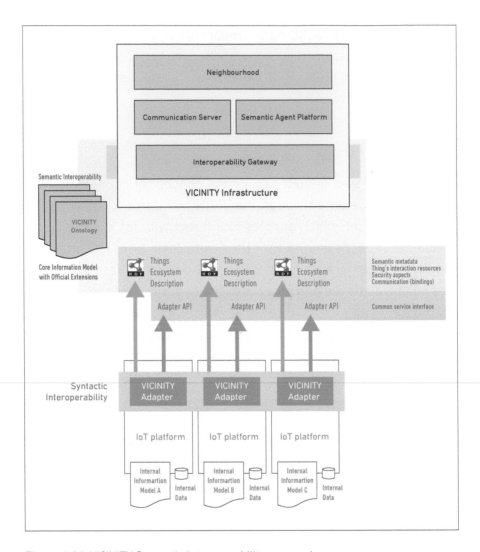

Figure 4.14. VICINITY Semantic interoperability approach

systems is integrated into VICINITY by its VICINITY Adapter through the VICINITY Gateway. Based on the setup of virtual neighbourhoods in the VICINITY Neighbourhood manager, VICINITY Adapters may access remote IoT objects based on semantic interoperability, for example a battery in an Energy ecosystem, and use them as a part of their ecosystem. Moreover, IoT objects shared by the VICINITY Adapter within a virtual neighbourhood may be accessed by value-added

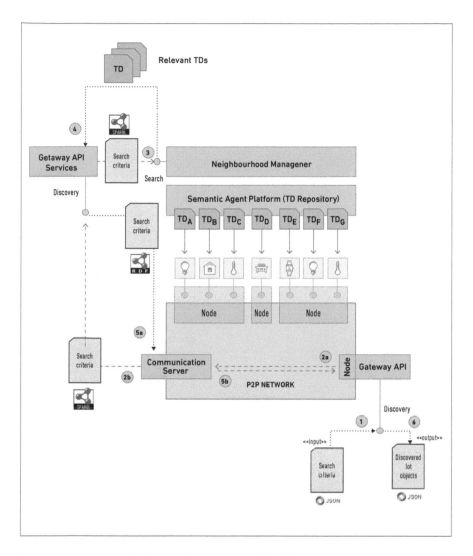

Figure 4.15. Semantic interoperability approach for Discovery

services to provide cross-domain services using a common VICINITY ontology.

One of the main challenges of implementing interoperability in the IoT context is to enable consumers to discover, in a distributed and dynamic scenario, those IoT objects that are relevant to their needs but without having any prior knowledge about them. The VICINITY cross-domain interoperability relies on:

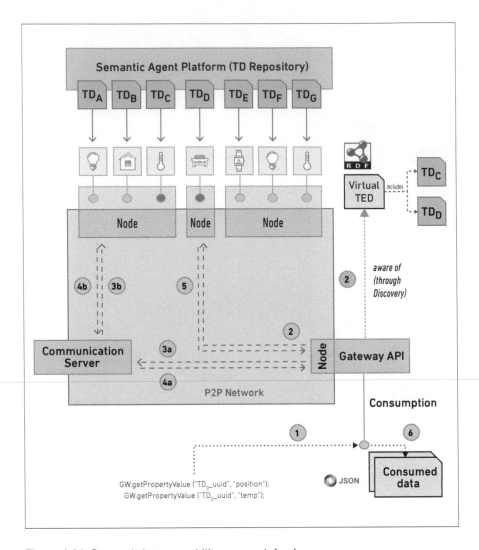

Figure 4.16. Semantic interoperability approach for Access

- The VICINITY ontology[19] as the common and abstract information model to be used;
- The Semantic agent platform as the semantic repository. The W3C Web of Things Thing Description (TD) is the framework to be used for describing any IoT object integrated in VICINITY;

---

19. http://vicinity.iot.linkeddata.es/

- Gateway Adapter APIs are the semantic mediators between the actual consumers, e.g., Adapters, and the repository of Thing Descriptions (TDs). Therefore, they provide an interface for discovery requests;
- Gateway Adapter APIs must be able to specify discovery needs as semantic-based search criteria (SPARQL query).

The first goal of semantic interoperability in VICINITY is to securely discover IoT objects. A discovery search initiated on a VICINITY Node through the Gateway API is performed in the Thing Description repository and constrained by the current setup virtual neighbourhood in the Neighbourhood Manager (discovered IoT objects are filtered by access rules defined on the virtual neighbourhood manager).

The second goal of semantic interoperability in VICINITY is to enable accessing to heterogeneous IoT objects in such a way that any of their interaction resources can be effectively *consumed* after they were discovered. Based on the result of the discovery process, IoT objects are accessed directly in the P2P network or through the Communication server services sending consumption request messages.

The Gateway API processes response messages separately and applies the corresponding access mappings specified in the previously given Virtual TED. For each response and after the data *lifting* process is completed, the Gateway API extracts the property value by querying the just lifted semantic data. Finally, the Gateway API returns each result obtained and represented in the common VICINITY format.

Used approach to semantic interoperability: Core Information Model with Extensions.

5

Advancing IoT platforms interoperability

The IoT-EPI efforts towards providing a framework for IoT platforms interoperability have already provided new advanced IoT platforms interoperability mechanisms and approaches. In this context, several interoperability gaps have been identified, that require further investigation and future research efforts.

Internet of Things (IoT) environments are rather complex with heterogeneous physical devices supporting various communication protocols, while they are possibly connected to an intermediary gateway and then to their virtual representations (i.e. services) running on different platforms. Thus, it is possible to interact with a single IoT device in many ways using its varied interfaces and representations.

IoT platforms require interoperability on multiple levels, which means finding the characteristic functionalities of each layer and defining meta-protocols that can be mapped on the ones used in the platforms (i.e. on the level of syntactic interoperability, the characteristic functionality is resource access). Resource access is realised through different protocols, for example O-MI, REST-based with JavaScript Online Notation (JSON) based on the W3C Web of Things (WoT) Description, OData-like and so on. While the functionality the protocols provide is basically the same (e.g. get/set/update/subscribe to a resource identified by some kind of identifier), the resource discovery functionality must be addressed across protocols.

Research on a layer-oriented approach is required to address providing tighter interoperability at all layers of IoT systems (device, network, middleware, application, data and semantics) with a strong focus on guaranteeing trust, privacy and security aspects within this interoperability. This interoperability approach also provides modules covering quality of service (QoS) and device management, service integration, external system services, storage and virtualisation.

Regarding semantic interoperability, current work focuses on defining and standardizing common vocabularies in given domains (e.g. iot.schema.org). Interesting direction is also the effort towards domain-agnostic aspects of any IoT object, following the WoT approach (with interaction patterns, links and security). However, standardization of models is not always a viable option. Therefore, research should also focus on techniques that enable semantic interoperability even if different information models are used like semantic mapping.

Layered approaches for interoperability allow the stakeholders or platform operators to select the best mechanism for interoperation. Management of such options provides coordination between layers, enhances cooperative solutions (e.g. gateways and network) and enables security management.

With regard to gateway interoperability, the inclusion of a programmable network layer based on software-defined networking (SDN)/network functions virtualisation (NFV) is critical for merging IoT and 5G and following the existing architectures. Using a programmable network has two advantages: management of mobility and management of QoS for massive IoT management.

The most promising aspects of a common framework for IoT interoperability are finding common approaches to resource access, resource discovery as well as semantic interoperability.

Further research is needed to address interoperability in the systems-of-systems view where all devices, 'things' and other information systems should be able to interoperate at the level of Internet protocols (TCP/IP, etc.) or even at the World Wide Web (WWW) level (HTTP, WebSockets, etc.). This signifies that lower-level protocols (Zigbee, LoRa, etc.) are abstracted away behind 'wrappers' and services using a limited set of IoT standards.

All IoT applications need to cope with the complex nature and sustainability requirements of interoperability in the IoT. For this, a framework is required for sustainable interoperability that especially targets the specific characteristics and constraints of the IoT.

A possible framework for sustainable interoperability in the IoT should be able to address the following aspects:

- Scalability management of interoperability in the IoT: to correctly support interoperability in the IoT, interoperability resources must be efficiently and effectively managed.
- Capacity of interoperability measurement in the IoT: to properly manage and execute interoperability in the IoT, a method must be reached to quantify and/or qualify the interoperability itself.
- Dynamic interoperability techniques and methodologies for the IoT: to achieve enduring interoperability in the complex IoT ecosystem, 'things' must be permitted to enter and dynamically interoperate without the need for heavy modification.

The validation aspect is very important in interoperability in general and even more so for the IoT. Testing and validating assert that interoperability methods, protocols and so on can cope with the specific nature and requirements of the IoT. We need to provide efficient and accurate test suites and an associated interoperability testing methodology that helps in testing thoroughly both the underlying protocols used by interconnected 'things', machines and smart objects, and the embedded services and applications. In this view, we have to consider that to most of the existing testing methods, interconnected 'resources' in the IoT are naturally distributed. As they are distributed, the usual and classical approach of a single centralised testing system dealing with all these components and the test execution is no longer applicable. The distributed nature of the tested components requires moving towards distributed testing methods.

The IoT research community has to agree on data sets for evaluating IoT platform ingestion performance, as well as agree on sets of real-time queries for evaluating an IoT platform digestion performance. Testing/experiment sequences have to be made explicit, transparent and consensus-based to ensure a fair and unbiased benchmarking process.

Implementation of the interoperability solutions should aim for an open source approach using different business models. The solutions need to be sustainable even after the end of the projects. Further research should focus on the key aspect of API homogenisation between platforms and federation authentication. This should allow for flexible cross-domain/cross-platform business process creation.

Utilising IoT (platform) technology for handling automated complex systems is a key challenge for the future. In this context, complex systems relate to systems that are heavily composed of rich sensing capabilities (e.g. radar, visual) and advanced actuation (e.g. controlling robot arms). The automation of these complex systems requires intelligent sensor data processing and reasoning to meaningfully control the actuation. IoT platform technology has the potential to facilitate and enhance the creation of systems significantly by being the interoperable interface within and between complex systems and the automation. Sensors, actuators and 'things' description formats, communication protocols and APIs defined by the IoT community need to use a lingua franca to enable such intelligent processing and reasoning.

Examples for such automated complex systems where IoT platform technology can provide a drastic development boost range from connected robots and robotic ecosystems over automated manufacturing lines that are integrated with logistic chains to automated vehicles (ships, aircrafts, trains, trucks, cars) and fleets.

Connecting components and systems to more complex systems is becoming possible with IoT platform technology. Thereby, intelligent and automated collaboration within and between such complex systems will be a central challenge. Automatically creating compositions of IoT components out of a pool of existing components (populated via a repository) will increase economic impact because reusing the components in multiple compositions will increase revenue created. Instead of controlling the composition centrally, it will be organised decentralised through intelligent choreographies.

In the future, complex services might even be created on the fly, for example when a car arrives in a new city and needs a specific service from the city's IoT ecosystem. How such context-based service composition can be implemented efficiently is one of the core research topics that requires further study.

Current efforts on semantic interoperability in the context of IoT focus on 'interoperability by standardization' trying to define standardized vocabularies to be used by multiple platforms and thus establishing semantic interoperability. However, there always will be scenarios where standardization is not desired or not even possible. To provide semantic interoperability in these cases other approaches to semantic interoperability are needed. The most prominent and promising is using data and query translation based on semantic mappings between the different information models used by the platforms. However, since the semantic mapping technologies are not trivial tools that would significantly make easier the process are required.

The unique identity of 'things' that spans across all platforms is still an unresolved issue. For instance, a car can be identified by a serial number or vehicle identification number (VIN), registration plate number, insurance number or some other identifier. The most appropriate one might depend on the usage context, so this feature should be a requirement for any IoT information system.

Although there are multiple ways of assigning identity to each individual 'thing' and each individual item, the work on aligning these different identity approaches and enabling their mapping to allow easy use of the same 'thing' in different environments has to continue.

The research work on a common IoT interoperability framework needs to be continued and requires a detailed and broad analysis of the different interoperability levels and their protocols paired with intensive scientific research as well as standardisation work.

A roadmap regarding interoperability is linked with the creation of the IoT ecosystems mainly with developers around the existing solutions (i.e. IoT-EPI). The functionalities of the interoperability solutions must be extended, including self-* mechanisms for controlling actuation, especially in future IoT autonomous systems.

6

References

---

[1] A. Gluhak, O. Vermessan, R. Bahr, F. Clari, T. Macchia, M. T. Delgado, A. Hoeer, F. Boesenberg, M. Senigalliesi and V. Barchetti, „Report on IoT platform activities," online at http://www.internet-of-things-research. eu/pdf/D03_01_WP03_H2020_UNIFY-IoT_Final.pdf, 2016.

[2] A. Velosa, Y. V. Natis, M. Pezzini, B. Lheureux and E. Goodness, *Gartner's Market Guide for IoT Platforms, 2015*, online at: https://www.gartner. com/doc/3086918/market-guide-iot-platforms.

[3] M. J. Perry, *Evaluating and Choosing an IoT Platform, 2016*, online at: https://www.thingworx.com/wp-content/uploads/WP_oreilly-media_ evaluating-and-choosing-an-iot-platform_978-1-491-95203-0_EN.pdf.

[4] *ThingWorx IoT Technology Platform*, online at: https://www.thingworx. com/platforms.

[5] L. Labs, online at: https://www.link-labs.com/blog/what-is-an-iot-plat-form.

[6] T. R. Gruber and et al., „A translation approach to portable ontology spe-cifications," *Knowledge acquisition*, 1993.

[7] A. Tolk and J. A. Muguira, „The levels of conceptual interoperability mo-del," in *Proceedings of the 2003 Fall Simulation Interoperability Workshop*, Citeseer, 2003, pp. 1-11.

[8] European Commission, European interoperability framework for pan-European egovernment services, ISBN 92-894-8389-X, 2011.

[9] H. van der Veer and A. Wiles, „Achieving Technical Interoperability – the ETSI Approach," *ETSI White Paper No.3, 3rd edition*, April 2008.

[10] A. Bröring, S. Schmid, C.-K. Schindhelm and A. Khelil, „Enabling IoT Eco-systems through Platform Interoperability," *IEEE Software, 34(1)*, pp. 54-61, 2017.

[11] S. Leminen, M. Westerlund, M. Rajahonka and R. Siurua, „Towards IoT ecosystems and business models," *Internet of Things, Smart Spaces, and Next Generation Networking*, pp. 15-26, 2012.

[12] *Merrian-Webster dictionary*, online at: https://www.merriam-webster.com/dictionary/semantics.

[13] T. Berners-Lee and L. Ora, „The semantic web," *Scientific american*, pp. 28-37, 2001.

[14] Network Centric Operations Industry Consortium, „Systems, capabilities, operations, programs, and enterprises (SCOPE) model for interoperability assessment," 2008.

[15] European Commission, „Reaping the full benefits of a digital single market," 2016.

[16] M. Jacoby, A. Antonic, K. Kreiner, R. Lapacz and J. Pielorz, „Semantic Interoperability as Key to IoT Platform Federation," *Interoperability and Open-Source Solutions for the Internet of Things*, Forthcoming 2017.

[17] S. Schmid, A. Bröring, D. Kramer, S. Kaebisch, A. Zappa, M. Lorenz, Y. Wang and L. Gioppo, „An Architecture for Interoperable IoT Ecosystems," in *Interoperability and Open-Source Solutions for the Internet of Things. InterOSS-IoT 2016*, Stuttgart, 2017.

[18] J. Hernandez-Serrano, J. Munoz, A. Bröring, O. Esparza, L. Mikkelsen, W. Schwarzott, O. Leon and J. Zibuschka, „On the Road to Secure and Privacy-preserving IoT Ecosystems," in *2nd International Workshop on Interoperability & Open Source Solutions for the Internet of Things (InterOSS-IoT 2016) at the 6th International Conference on the Internet of Things (IoT 2016)*, Stuttgart, Germany, 2017.

[19] A. S. Thuluva, A. Bröring, G. P. Medagoda Hettige Don and D. Anicic, „Recipes for the Semantic Composition of IoT Ingredients," in *7th International Conference on the Internet of Things (IoT 2017)*, Linz, Austria, 2017.

[20] W. Schladofsky, J. Mitic, A. Megner, C. Simonato, L. Gioppo, D. Leonardos and A. Bröring, „Business Models for Interoperable IoT Ecosystems," in *Interoperability and Open-Source Solutions for the Internet of Things. InterOSS-IoT 2016*, Stuttgart, Germany, 2017.

[21] W. W. W. C. (W3C), *Web of Things at W3C, W3C Begins Standards Work on Web of Things to Reduce IoT Fragmentation*, online at: https://www.w3.org/WoT/.

[22] T. O. Group, *An Introduction to Internet of Things (IoT) and Lifecycle Management, Maximizing Boundaryless Information Flow through Whole-of-Life Lifecycle Management Across IoT*, online at: https://www2.opengroup.org/ogsys/catalog/W167.

[23] Främling, Kary, Kubler, Sylvain, Buda and Andrea, „Universal Messaging Standards for the IoT from a Lifecycle Management Perspective.," *IEEE Internet of Things*, Bd. 1, Nr. 4, pp. 319-327, 2014.

[24] I. Podnar Zarko et al, „Towards an IoT framework for semantic and organizational interoperability," in *Global Internet of Things Summit (GIoTS)*, Geneva, 2017.

[25] S. Sciancalepore, M. Pilc, S. Schrder, G. Bianchi, G. Boggia, P. M., P. G. M. Plocielnik and H. Weisgrab, „Atribute-based access control scheme in federated IoT platforms," *LNCS 10218: Interoperability and Open-Source Solutions for the Internet of Things*, pp. 123-138, 2017.

[26] IERC, „IoT Semantic Interoperability: Research Challenges, Best Practices, Recommendations and Next Steps. Position Paper March 2015".

[27] Consorzio per il Sistema Informativo (CSI-Piemonte), online at: http://www.csipiemonte.it/web/en/.

[28] Wubby, online at: http://www.wubby.io/.

[29] OpenIoT Consortium, online at: http://www.openiot.eu/.

[30] Verkehrsinformationszentrale Berlin (VIZ), online at: https://viz.berlin.de/.

[31] World Sensing, online at: http://www.worldsensing.com/.

[32] *Amazon AWS IoT*, Online at: http://docs.aws.amazon.com/iot/latest/developerguide/what-is-aws-iot.html.

[33] M. Ganzha, M. Paprzycki, W. Pawłowski and P. Szmeja, „Towards common vocabulary for IoT ecosystems—preliminary considerations," *Intelligent Information and Database Systems - 9th Asian Conference, ACIIDS*, 3-5 April 2017.

[34] M. Ganzha, M. Paprzycki, W. Pawlowski, P. Szmeja and K. Wasielewska, „Semantic Technologies for the IoT-An INTER-IoT Perspective," *Internet-of-Things Design and Implementation (IoTDI), 2016 IEEE First International Conference*, pp. 271-276, April 2016.

[35] K. Vandikas and V. Tsiatsis, „Performance Evaluation of an IoT Platform," *Eighth International Conference on Next Generation Mobile Apps, Services and Technologies, IEEE*, p. 141–146, 2014.

[36] F. Ramparany, F. Marquez, J. Soriano and T. Elsaleh, „Handling smart environment devices, data and services at the semantic level with the FI-WARE core platform," *IEEE International Conference on Big Data (Big Data)*, pp. 14-20, 2014.

[37] M. Serrano, H. Nguyen Mau Quoc, D. Le Phuoc, M. Hauswirth, J. Soldatos, N. Kefalakis, P. Prakash Jayaraman and A. Zaslavsky, „Defining the Stack for Service Delivery Models and Interoperability in the Internet of

Things: A Practical Case With OpenIoT-VDK," *III Journal on Selected Areas in Communications*, 2015.

[38] BETaaS Consortium, http://www.betaas.eu.

[39] BETaaS, *Building the Environment for the Things as a Service, TaaA Reference Model, 2013*, online at: http://www.betaas.eu/docs/deliverables/BETaaS%20-%20D1.4.2%20-%20TaaS%20Reference%20Model%20v1.0.pdf.

[40] Nextworks, *Symphony, the harmony of living*, online at: http://www.hello-symphony.it/eng/home-eng.

[41] FIWARE, *FIWARE Data Models*, online at: https://www.fiware.org/data-models/.

[42] FIWARE, *FIWARE Accelerator Programme*, online at: https://www.fiware.org/fiware-accelerator-programme/.

[43] W. W. W. C. (W3C), *Resource Description Framework (RDF)*, online at: https://www.w3.org/RDF/.

[44] W. W. W. C. (W3C), *SPARQL 1.1 Overview, W3C Recommendation 21 March 2013*, online at: https://www.w3.org/TR/sparql11-overview/.

[45] W. W. W. C. (W3C), *W3C Quality Assurance*, online at: https://www.w3.org/QA/Test/.

[46] O. G. C. (OGC), online at: https://cite.opengeospatial.org/.

[47] H. Stuckenschmidt, C. Parent and S. Spaccapietra, *Modular ontologies: concepts, theories and techniques for knowledge modularization, (Vol.5445) Springer, 2009*, online at: http://sociotal.eu/.

[48] FIWARE, *FIWARE Architecture*, online at: https://forge.fiware.org/plugins/mediawiki/wiki/fiware/index.php/FIWARE_Architecture.

[49] *Microsoft Azure Documentation Reference Architecture*, online at: https://docs.microsoft.com/en-us/azure/architecture/reference-architectures/.

[50] *Transaction Processing Performance Council (TPC)*, online at: http://www.tpc.org/.

[51] *Predix Architecture*, online at: https://www.predix.com/sites/default/files/ge-predix-architecture-r092615.pdf.

[52] *Tibbo Agregate IoT Integration platform*, online at: http://aggregate.tibbo.com/.

[53] A. Foster, *Enhanced data storage capabilities for IBM Watson IoT Platform*, online at: https://developer.ibm.com/iotplatform/2016/07/25/enhanced-data-storage-capabilities-for-ibm-watson-iot-platform/.

Authors

Arne Bröring, Siemens, Germany.

Achille Zappa, Insight Centre for Data Analytics, National University of Ireland Galway, Ireland.

Ovidiu Vermesan, SINTEF, Norway.

Kary Främling, Aalto University, Finland.

Arkady Zaslavsky, CSIRO, Australia.

Regel Gonzalez-Usach, Universitat Politecnica de Valencia (UPV), Spain.

Pawel Szmeja, Systems Research Institute, Polish Academy of Sciences (SRI-PAS), Poland.

Carlos E. Palau, Universitat Politecnica de Valencia (UPV), Spain.

Michael Jacoby, Fraunhofer IOSB, Germany.

Ivana Podnar Zarko, Univ. of Zagreb, Faculty of Electrical Engineering and Computing, Croatia.

Sergios Soursos, Intracom Telecom, Greece.

Corinna Schmitt, University of Zurich, Department of Informatics, Switzerland.

Marcin Plociennik, PSNC, IBCh PAS, Poland.

Srdjan Krco, DunavNET, Serbia.

Stylianos Georgoulas, University of Surrey, UK.

Iker Larizgoitia, Evrythng, UK.

Nenad Gligoric, DunavNET, Serbia.

Raúl García-Castro, Universidad Politécnica de Madrid, Spain.

Fernando Serena, Universidad Politécnica de Madrid, Spain.

Viktor Oravec, bAvenir, Slovakia.

Raffaele Giaffreda, FBK CREATE-NET, Italy.

Csaba Kiraly. FBK CREATE-NET, Italy.

Contributors

Stefan Schmid, Bosch Corporate Research, Germany.

Corina-Kim Schindhelm, Siemens, Germany.

Abdelmajid Khelil, Landshut University of Applied Sciences, Germany.

Sebastian Käbisch, Siemens, Germany.

Denis Kramer, Bosch Software Innovations, Germany.

Danh Le Phuoc, Technical University of Berlin, Germany.

Jelena Mitic and Darko Anicic, Siemens, Germany.

Ernest Teniente, Universitat Politècnica de Catalunya, Spain.

Robert Hellbach, BIBA, Germany.

Sylvain Kubler, University of Luxembourg, Luxembourg.

Jérémy Robert, University of Luxembourg, Luxembourg.

Johan Schabbink, Neways Electronics, Netherlands.

João Garcia, Ubiwhere, Portugal.

Ricardo Vitorino, Ubiwhere, Portugal.

Gerhard Duennebeil, AIT, Austria.

Matteo Pardi, Nextworks, Italy.

Michele Bertolacci, Navigo, Italy.

Raquel Ventura Miravet, S&C, Spain.

Elena Garrido Ostermann, ATOS, Spain.

Aleksandar Antonic, University of Zagreb, Croatia.

Riccardo Pozza, University of Surrey, UK.

Pekka Liedes, UPCode, Finland.

Colin O'Reilly, University of Surrey, UK.

Liisa Hakola, VTT, Finland.

Mick Wilson, Fujitsu, UK.

Roy Bahr, SINTEF, Norway.

Rafał Tkaczyk, SRIPAS, Poland.

Clara Valero, UPV, Spain.

7

Annex

The following section captures the different platforms the IoT-EPI projects are utilizing in their project.

The purpose is to present the current state of play and to get a nover view of for the diversity and overlap of IoT platforms across the seven emerging ecosystems.

The analysis briefly introduces the IoT platform name, whether the platform is commercial or not and provides a brief description of its main purpose.

AGILE		
IoT platform	**Nature of platform**	**Brief description**
Resin.io	Commercial	Device management platform for Linux based IoT devices. It makes it simple to deploy, update, and maintain code running on remote devices.
Eclipse IoT	Open source	Eclipse IoT provides the technology needed to build IoT Devices, Gateways, and Cloud Platforms. It also provides open source implementations for IoT standards such as MQTT, CoAP, LWM2M, OneM2M, OPC-UA and more.
NodeRED	Open source	Tool for wiring together hardware devices, APIs and online services in new and interesting way. IoT service enablement platform developed by IBM.

BIG IoT		
IoT platform	**Nature of platform**	**Brief description**
Smart Data Platform	Open source	CSI Piemonte's Smart Data Platform (SDP) is a self-service platform enabling application development based on Internet of Things and Big Data [27]. SDP is based on project Yucca which allows for interconnecting applications, social networks, systems and distributed objects and collecting data and information, by processing and analysing them to develop end-to-end solutions
Smart City Platform	Commercial, Bosch	Considering solutions for Smart Cities, the requirements differ from those known for classical enterprise applications. In fact, Smart City installations are composed of many different solutions individually customized for the city, but with a common need w.r.t. operation, data sharing and security. The Smart City platform (SCP) targets to connect the silos in the Smart City, i.e., governance, mobility, energy, environment, industry life, tourism, etc. Bosch SCP offers tools and methods to develop, operate and maintain such systems without sacrificing data security and privacy.
Wubby Platform	Commercial, Econais	Wubby is an ecosystem of software components and services for rapid development of everyday objects [28]. Everyday objects are physical objects embedded with electronics, software, sensors and network connectivity to collect and exchange data.
OpenIoT Platform	Open Source	OpenIoT is a sensor middleware platform that eases the collection of data from heterogeneous sensors, while ensuring their semantic annotations [29]. It enables semantic interoperability in the cloud and provides IoT app development tools.

Traffic Information Centre Platform	Commercial, VMZ	The TIC mobility platform provided by VMZ is a data and service platform that has been developed to provide comprehensive information on all mobility options available in Berlin [30]. The platform includes real-time data from the traffic information center, mobility operators and infrastructure providers and provides a multimodal routing platform using the modal router offered by third parties.
Bitcarrier/ Sensefield/ FastPrk	Commercial, World Sensing	Worldsensing provides a unique traffic management portfolio for Smart Cities that includes Bitcarrier, a real-time intelligent traffic management and information solution designed for both road and urban environments [31]. Fastprk provides an intelligent parking system and Sensefields provides an innovative system for detecting and monitoring vehicles and traffic flow.
BEZIRK Platform	Open Source	Bosch's Bezirk platform is a peer-to-peer IoT middleware for both communication and service execution on local devices following the service-oriented paradigm. Bezirk is developed with a view to facilitate asynchronous interactions between the different components of an application with respect to distribution across different devices in a network.

BioTope

IoT platform	Nature of platform	Brief description
O-MI/O-DF Reference Implementation	Open source	Implementation of O-MI and O-DF standards for the IoT that makes it easy to set up standard-based IoT node instances. Mainly used for "sandbox" installations but can be scaled up for "industry-level" purposes.
DIALOG	Open source	IoT Middleware originally developed by Aalto in 2001, which has been further developed and used in numerous research projects as well as industrial pilots.
NodeRED	Open source	Tool for wiring together hardware devices, APIs and online services in new and interesting way. Visual IoT service enablement platform developed by IBM.
Warp 10	Open source	Platform for storage, management and analysis of IoT data, especially for Geo Time Series.
FIWARE	Open source	FIWARE is a middleware platform for the development and global deployment of applications for Future Internet. It is an outcome of a large investment of the EU into large-scale research programme involving network vendors and operators.
Open IoT	Open source	OpenIoT is a sensor middleware platform that eases the collection of data from heterogeneous sensors, while ensuring their semantic annotations. It enables semantic interoperability in the cloud and provides IoT app development tools.
Mist	Closed-source	Software stack for distributed, secure IoT deployments of ControlThings.
eAir web	Closed-source	Cloud service for remote use and management of Enervent Air Handling units.

| Other | Open/Closedsource | Numerous platforms such as BMW's platform, several Smart Parking platforms in Helsinki, OpenDataSoft's platform. Estimated over 10 different platforms used now or in the future. |

INTER-IoT		
IoT platform	**Nature of platform**	**Brief description**
SEAMS	EU project	Smart, Energy-Efficient and Adaptive Management Platform (SEAMS) is a state-of-the art prototype-monitoring tool developed and implemented within the framework of the European project SEA TERMINALS at Noatum Container Terminal Valencia. The SEAMS platform prototype is capable of monitoring the machines and equipment that are being used at a Port Container Terminal.
I3WSN	Academic platform	Industrial Intelligent Wireless Sensor Networks for indoor environments, platform developed by Universitat Politécnica de Valencia
Unical BodyCloud	Open source	BodyCloud is an open platform for the integration of BSNs with a Cloud Platform-as-a-Service (PaaS) infrastructure and it's currently based on Google App Engine.
NodeRED	Open source	Tool for wiring together hardware devices, APIs and online services in new and interesting way. We will use it in the AS2AS interoperability framework
OpenIoT	Open source	OpenIoT is a sensor middleware platform that eases the collection of data from heterogeneous sensors, while ensuring their semantic annotations. It enables semantic interoperability in the cloud and provides IoT app development tools.
FIWARE	Open source	FIWARE is a middleware platform for the development and global deployment of applications for Future Internet. It is an outcome of a large investment of the EU into large-scale research programme involving network vendors and operators.

UniversAAL	**Open Source**	UniversAAL is an IoT platform developed in the framework of an FP7 project and applied currently in different AAL, eHealth and AHA environments.
Eclipse OM2M	**Open Source**	The Eclipse OM2M project, initiated by LAAS-CNRS, is an open source implementation of oneM2M and SmartM2M standard.
WSO2	**Open Source ASL2.0**	WSO2 IoT Server enables device manufacturers and enterprises to connect and manage their devices, build apps, manage events, secure devices and data, and visualize sensor data in a scalable manner.
Microsoft Azure IoT Suite	**Proprietary Microsoft**	Microsoft Azure is a full cloud-based platform with IoT specific components to support connection of devices to the cloud, analyse, store and visualize captured data. Microsoft Azure Suite provides an easy to configure back-end for IoT deployments. It is domain agnostic and provides no models for data. It can be combined with advanced data analytics, machine learning and other components.
Amazon AWS IoT	**Proprietary Amazon**	AWS module specially intended to IoT systems [32]. It enables a straightforward access to Amazon Cloud thanks to an easy to use management interface and a REST API to control the status of the things connected. Once data is sent to the AWS IoT, then it can be used the huge ecosystem of AWS cloud solutions. This platform is completely domain agnostic and provides a strong security protection.

TagItSmart		
IoT platform	**Nature of platform**	**Brief description**
SocIoTal	Open source	Community IoT platform with privacy aware data sharing. Developed by the FP7 SOCIATAL project.
FIWARE	Open Source	FIWARE is a middleware platform for the development and global deployment of applications for Future Internet. It is an outcome of a large investment of the EU into large-scale research programme involving many network vendors and operators.
EVRYTHNG	Commercial	IoT Smart product platforms. The platform collects, manages and applies real-time data from smart products and smart packaging to drive IoT applications.
RunMyProcess	Commercial	Build device independent, connected applications with strong business process integration; Deploy systems at global scale; Run secure, reliable and scalable operations. Thousands of pre-built connectors to quickly integrate IoT-enabled devices, cloud services, and social media with on premise enterprise applications and systems.
Microsoft Azure	Commercial	Full cloud-based platform with IoT specific components to support connection of devices to the cloud, analyse, store and visualize captured data. Can be combined with advanced data analytics, machine learning and other components.

symbioTe		
IoT platform	**Nature of platform**	**Brief description**
OpenIoT	**Open source**	OpenIoT is a sensor middleware platform that eases the collection of data from heterogeneous sensors, while ensuring their semantic annotations. It enables semantic interoperability in the cloud and provides IoT app development tools.
Symphony	**Commercial, Nextworks**	Networks platform for the integration of home and building control systems. Symphony can monitor, supervise and control many different building systems, devices, controllers and networks available from third-party suppliers. It is a service-oriented middleware, able to integrate several functional subsystems into a unified IP based platform.
Mobility Back-end as a Service (MoBaaS)	**Commercial, Ubiwhere**	System integration platform to wrap around different city data sources. Application enablement environment geared towards smart city apps focusing on transport and mobility aspects of cities.
nAssist	**Commercial, Sensing and Control Systems S.L.**	A software platform designed and conceived to allow agile, continuous management of data in the fields of energy efficiency, security and automation. Cloud-based communication software that enables clients to easily and intelligently connect machines and devices to the cloud and then process, transform, organize and store machine and sensor data.
Navigo Digitale IoT platform	**Commercial, Navigo**	A vertical IoT platform created to manage digital assets pertaining to harbours used for boating and yachting. Its focus is to provide services to the harbour's activities (B2B) and to its end-users (B2C).

| KIOLA | Commercial, AIT | A mobile health data collection and online therapy management system. It integrates different sensor devices on the client side and provides backend interfaces for health management systems. |

VICINITY		
IoT platform	**Nature of platform**	**Brief description**
LinkSmart	Open source	IoT middleware originally developed in the Hydra project. It allows developers to Incorporate heterogeneous physical devices into their applications through easy-to-use web services for controlling any device.
IoTivity	Open Source	IoTivity is an open source software framework enabling seamless device-to-device connectivity to address the emerging needs of the Internet of Things.
SiteWhere	Open Source	The Open Platform for the Internet of Things provides a set of APIs to connect devices through MQTT, AMQP, Stomp and other protocols, enabling self-registration of devices and manipulation with devices using batches.
Eclipse Kura	Open Source	Eclipse Kura is Java/OSGi-based framework for IoT gateways including a set of APIs to access to hardware in/outputs, management of network configurations, communication with M2M/IoT Integration Platforms. Moreover, Eclipse Kura supports Apache Camel routes to introduce business logic on the level of the IoT gateway.
TinyMesh	Commercial	Self-managed and self-healing mesh network platform for TinyMesh compatible devices that enables the collection of measures and performance in the network.
Gorenje Cloud services	Commercial	Cloud based IoT platform to control and maintain Gorenje household appliances.

List of Figures and Tables